CW00703536

Jack Bowyer DipArch FRIBA is a practising architect and also a lecturer at Croydon Technical College. He is the author of several books, including *Guide to Domestic Building Surveys*, *Building Technology Books 1, 2 and 3*, *Practical Specification Writing*, and *Evolution of Church Building*.

Architectural Press Legal Guides

Small Works Contract Documentation

and how to administer it

Third edition

Jack Bowyer DiplArch (Leeds)

The Architectural Press: London

First published in 1976 by the Architectural Press Ltd,
9 Queen Anne's Gate, London SW1 9BY
Second edition 1978
Third edition 1986

British Library Cataloguing in Publication Data

Bowyer, Jack
Small works contract documentation: and how to administer it. – 3rd ed. – (Architectural Press legal guides)
1. Building – Contracts and specifications – Great Britain
I. Title
692'.8'0941 TH425

ISBN 0-85139-976-2

The author and publishers wish to record their thanks to Robert Johnstone, Legal Adviser to the Royal Institute of British Architects, for his advice on chapter 5

The specimen RIBA forms are reproduced by permission of RIBA Publications Ltd, the copyright holder.

The contracts referred to in the text are:

Standard Form of Building Contract 1980 edition
Intermediate Form of Building Contract 1984
Agreement for Minor Building Works first issued 1980: October 1981 reprint issued by the Joint Contracts Tribunal

By the same author:

Guide to Domestic Building Surveys
A History of Building (Orion Books)
With J. Trill, *Problems in Building Construction*
Small Works Supervision
Evolution of Church Building (Granada)
Building Technology Books 1, 2 and 3 (Newnes Butterworth)
Practical Specification Writing (Hutchinson)

Typeset by Crawley Composition Ltd
Printed in Great Britain by Biddles Ltd, Guildford and King's Lynn

Contents

Introduction xi

1 Inception of commission to invitation to tender 1

Letter from architect to client: acknowledgement of receipt of commission 2
Letter from architect to client: confirmation of instructions and Conditions of Engagement 3
Specimen Memorandum of Agreement 5-8
Letter from architect to client: to enclose sketch plans and approximate estimate 9
Letter from architect to client: acknowledgement of approval of sketch proposals 11
Letter from architect to client: appointment of consultants 12
Letter from architect to consultant: confirmation of nomination 12
Letter from architect to sub-contractors: preliminary information 14
Letter from architect to local planning authority: application for approval 15
Letter from architect to client: to enclose copies of final production drawings 16
Letter from architect to local authority building control: application for approval 17
Letter from architect to quantity surveyor: preparation of bills of quantities 19
Letter from architect to prospective tenderers: preliminary invitation to tender 21-22
Letter from architect to prospective tenderers: invitation to tender 23
Example of Form of Tender 24-25
Example of Schedule of Rates 26
Letter from architect to client: to advise tenders invited 27

2 Placing the contract 29
Letter from architect to quantity surveyor: to check tenders 28
Letter from architect to all tenderers: acknowledgement of receipt of tender 28
Letter from architect to tenderer: return of late tender unopened 31
Letter from architect to client: report on tenders 31
Letter from architect to unsuccessful contractor submitting lowest tender: to advise that the tender has not been accepted by client 33
Letter from architect to client: confirmation of written instructions to accept a tender 33
Letter from architect to client: confirmation of verbal instructions to accept a tender 34
Letter from architect to successful contractor: acceptance of tender 34

2.1 The Standard Form of Contract 35

2.2 The Intermediate Form of Contract 35

2.3 The Agreement for Minor Building Works 36
Letter from architect to successful contractor: to enclose contract documents for signature and return 37
Letter from architect to client: to seek appointment for signature of contract documents 38
Letter from architect to contractor: to advise completion of contract documents by employer 38
Letter from architect to unsuccessful tenderers: to advise that their tender is unsuccessful 40

3 The contract I: general 41
Letter from architect to contractor: to arrange initial site meeting 42
Report: site meeting 43
Letter from architect to employer: insurance of works 44
Letter from architect to contractor: insurance of works 45
Specimen Architect's Instruction 47
Specimen Clerk of Works' Direction 48
Letter from architect to contractor: sub-letting 50

Letter from contractor to architect: datum levels and setting out of works 50
Letter from architect to contractor: datum levels and discrepancy of dimensions 51
Letter from architect to contractor: discrepancy between contract drawings and specification 53
Architect's Instruction, from architect to contractor: discovery of antiquities 53
Specimen Standard form of Tender for Nominated Sub-contractors NSC/1 55-66
Specimen Form of Agreement between Employer and Nominated Sub-contractor NSC/2 67-70
Specimen Form of Employer/Specialist Agreement ESA/1 72-77
Architect's Instruction, from architect to contractor: instruction regarding prime cost sums 79
Specimen Standard Form of Employer/Nominated Sub-contractor Agreement NSC/2a 80-83
Specimen Standard form NSC/3 for Nomination to Sub-contractors to be used with NSC/1 84
Architect's Instruction, from architect to contractor: instruction regarding prime cost or provisional sums 85
Specimen Form of Tender for use by nominated suppliers TNS/1 86-89
Architect's Instruction, from architect to contractor: removal of material or goods not in accordance with the contract 90
Specimen Form of Warranty to be given by Nominated Supplier TNS/2 91-92
Architect's Instruction, from architect to contractor: statutory obligations 93
Specimen Interim Certificate and VAT provision 95
Specimen Certificate of Interim/Progress Payment 97
Specimen Authenticated Receipt 98-99
Specimen Statement of Retention and of Nominated Sub-contractor's values 101
Specimen Notification to Nominated Sub-contractor concerning Amount included in Certificate 102
Architect's Instruction, from architect to contractor: open up works for inspection or testing 104
Architect's Instruction, from architect to contractor: restoration and repair after fire 105
Architect's Instruction, from architect to contractor: postponement of work 106
Specimen Notification of an Extension of Time 108
Specimen Notification of Revision to Completion Date 110

Notice of Nominated Sub-contractor's Non-completion 111
Notice of Non-compliance 112

4 The contract II: determination 113
Letter from architect to contractor: notice specifying default 116
Letter from employer to contractor: determination by employer 117
Letter from architect to contractor: assignment of agreements 119
Letter from architect to contractor: removal of materials from site 119
Letter from architect to contractor: certification of amount of expenses in completion after determination 121

5 The contract III: arbitration 123
Letter from architect to RIBA: request for Form A relating to appointment of an arbitrator 125
Specimen Form for Appointment of an Arbitrator 127
Letter from architect to president RIBA: covering return of Form A 128
Letter from architect to contractor: asking for agreement to appointment of named arbitrator 129

6 The contract IV: the defects liability period 133
Specimen Certificate of Partial Possession 135
Specimen Certificate of Practical Completion (or by Nominated Sub-contractor) 136
Specimen Certificate of Practical Completion 137
Specimen Certificate of Non-completion 138
Letter from architect to contractor: certification of practical completion, Agreement for Minor Building Works 139
Notice of Delay in Completion 140
Architect's Instruction, from architect to contractor: to make good a defect within the Defects Liability Period 141
Letter from architect to contractor: to enclose Schedule of Defects 142
Schedule of Defects 143
Letter from architect to contractor: works extra to contract 144
Specimen Certificate of Making Good Defects 145

7 The contract V: the final account 147

Method A 149-152
Method B 154
Letter from architect to contractor: agreement to Final Account 155
Final Account 156
Letter from architect to client: to arrange meeting regarding Final Account 157
Letter from architect to client: agreement to Final Account 157
Specimen Final Certificate 158
Letter from architect to contractor: issue of Final Certificate 159
Letter from architect to client: to enclose statement of fees and expenses 159
Letter from architect to contractor: expression of thanks 159

Glossary of terms 160

Introduction

A large part of an architect's working life is spent on the administration of contracts. Whether these contracts are large or small, a continuous and ever-growing volume of paperwork is required to deal with the everyday problems they bring. Apart from the essentials of contract law, the architect receives little training in contract administration and, unless he is very fortunate, he can easily become overwhelmed by its complexities.

Small works may range from the construction of a simple factory unit with production shed and administrative wing, to the renovation and rehabilitation of a large country house. The variety covered by small works contracts is therefore very wide, requiring a high degree of skill and competence in their documentation.

This state of affairs is becoming even more apparent with the general use of three differing forms of contract, the Standard Form of Building Contract 1980 edition (generally known as JCT 80), the Intermediate Form of Building Contract (IFC 84), and the Agreement for Minor Building Works (MW 80). Generally, the Standard Form (in both With and Without Quantities forms) is used where the size, extended contract period or complexity of the works are such as to need contractual conditions: these conditions cover a large variety of likely situations, all of which need to be defined and evaluated. The Agreement for Minor Building Works, on the other hand, is a simple document providing for the minimum of eventualities and is employed for relatively simple contracts of short duration. Totally different in format, each requires its own methods of administration.

This book seeks to explain the varied problems and complexities of small works administration, from the inception of the commission to the issue of the Final Certificate. Both contract forms are considered, and their differing requirements are explained and illustrated with examples.

The first two chapters cover pre-contract activities, Conditions of Engagement, and the delicate matter of fees, defining meanwhile the relationship between architect and client. The submission of drawings for approval, the appointment and servicing of consultants, and the invitation to tender are also included in this section.

The administration and documentation of the contract are fully described in the first three chapters. Chapter 3 deals with those general matters which form part of all contracts: instructions, placing of sub-contracts, the issue of certificates. Chapter 4 covers the problems of determination, by both the contractor and the employer. The last two chapters are concerned with contract administration, from practical completion to Final Account, and offer suggestions on the presentation of the account for the client. The text ends with the issue of the Final Certificate.

An important feature of this book is the commentary on each section of the text, explaining why the action has to be taken and, where necessary, providing alternative methods for dealing with the problem. In addition, many of the standard forms and certificates published by RIBA Publications Ltd have been reproduced and inserted in the text to illustrate examples.

1 Inception of commission to invitation to tender

Invitations to carry out professional commissions arrive in many forms. In most cases a telephone call is made, in which the prospective client outlines his requirements and receives his selected architect's verbal acceptance of the commission. Sometimes the invitation is made at a social or business meeting. More rarely, in the case of a local authority or similar body, a formal letter is the first intimation of appointment.

In whatever form the invitation arrives, acceptance and any meeting or future arrangements should be confirmed immediately, in writing (see Fig 1.1). This indicates a businesslike approach which will immediately appeal to the client.

At this stage the following administrative actions should be put in motion:

o A file should be opened and a job number or job reference provided for the commission. This job number or reference should be quoted on all drawings, letters, or other documents which deal with any matter in connection with the particular commission.*

o A simple description of the proposed works should be agreed and this, together with the address of the site, should be used as a job description throughout the commission.

Always make sure that the writer's reference is added to all letters, usually in the form of initials together with those of the typist/secretary.

All letters and documents should be dated.

Written confirmation of instructions received at the site meeting should be sent to the client as soon as possible (see Fig 1.2). In some cases the client will send a detailed brief and sketches indicating the manner in which he sees his requirements related to the site. Usually it will be necessary to draw this information from him by a process of question and suggestion. In this event the matters agreed should be confirmed as minutes of the meeting, to avoid any possibility of confusion over the exact requirements.

*References, dates and job descriptions are omitted from future examples for simplification

Fig 1.1 Letter from architect to client: acknowledgement of receipt of commission

IJ/SP [*Writer's reference*] 1 January 198- [*Date of writing*]

Dear Sir

re proposed new building, South Road, New Town
Job Number 259

Thank you for your invitation to act as your architect in connection with your proposed new building in South Road, New Town. I shall be very happy to act for you and will meet you on site as you suggest on [*insert date*] at [*insert time*] to discuss this matter further.

Yours faithfully

Fig 1.2 Letter from architect to client: confirmation of instructions and Conditions of Engagement

Dear Sir

With reference to our meeting on site on [*insert date*], I should like to acknowledge receipt of your sketches and typed brief describing in detail the requirements of your company for your new building. I will arrange for a measured survey of the site to be prepared and levels taken. As soon as possible after this I will send you sketch proposals on the lines of our discussion.

I confirm agreement in respect of the services my firm are to carry out on your behalf, your agreement to the conditions of my appointment and fees and expenses applicable to my professional services. These have been incorporated in a formal Memorandum of Agreement, one copy being signed for and on behalf of my firm for your use and retention and one blank for your signature and return to me at your earliest convenience.

Yours faithfully

It is usually unwise to commit oneself too emphatically at an initial site meeting until the requirements of the brief and the site conditions are fully appreciated. Keep the discussion to generalities.
It is certain that the matter of fees will be raised. It may well be that your client is fully aware of the details of the appointment of architects, their conditions of engagement and fee scales. Usually, he will be ignorant of the details and in most cases he will not appreciate the conditions under which an Architect's Appointment is made. The question of fees must be settled at this stage as the fees and expenses quoted are recommended only and not mandatory. The booklet *Architect's Appointment – Small Works* (RIBA Publications) gives guidance on these matters. When fee scales and expenses have been agreed and the client fully advised of the implications of the details of architects' services and conditions of appointment a copy of the RIBA Memorandum of Agreement (see pp.5–8) should be prepared and completed to cover the agreed details. The Memorandum should be signed by both parties, the original being kept by the client with a copy in the project file for your reference. Alternatively, two forms can be prepared, one being signed by each and copies exchanged. This method of formalizing contractual matters between architect and client are applicable to contracts under JCT 80, IFC 84 and the Agreement for Minor Building Works. These conditions change at intervals and you must have a copy of those relevant to your particular commission. If further work follows and the conditions or fee scales change, you can then issue an up to date copy without any problems or embarrassment. Do not forget to point out to your client that VAT will be added to your fees and expenses under the Finance Act, 1972 if you are registered with the Customs and Excise.
Sketch plans should be sent to the client as soon as possible after instructions have been received (see Fig 1.3). Any long delay could indicate a lack of interest in the commission. Two copies of the sketches should be sent, one for the client to keep and one on which he can indicate amendments for consideration. Laymen find it very difficult to explain their requirements precisely in words, but can generally amend a drawing. In fact, in the majority of cases, their requirements are both vague and unpractical; the production of a sketch plan by the architect will put them into a practical and precise form.
The brief may be found to be impracticable through a physical cause discovered during the preparation of the sketch. For example: dimensions of site vary from information given; or sewer depths are too shallow to accommodate long runs of drain on site, necessitating re-planning of sanitary accommodation; or levels are unpractical for proposals envisaged; or access is shifted by the local authority, etc. In cases like this, the amendments and the reasons for them should be

Memorandum of agreement

between Client and Architect for use with the RIBA Architect's Appointment

This Agreement

is made on the day of 19

between
(insert name of client)

of

(hereinafter called the 'Client')

and
(insert name of architect or firm of architects)

of

(hereinafter called the 'Architect')

Now it is hereby agreed

that upon the Conditions in Parts 3 and 4 of the Architect's Appointment
(1982 Edition Revision), a copy of which is attached hereto,

save as excepted or varied by the parties hereto in the attached Schedule of Services and Fees, hereinafter called the 'Schedule',

and subject to any special conditions set out or referred to in the Schedule:

1 the Architect will perform for the Client the services listed in the Schedule in respect of

(insert general description of project)

at
(insert location of project)

2 the Client will pay the Architect on demand for the services, fees and expenses indicated in the Schedule;

3 consultants will be appointed as indicated in the Schedule;

4 site staff will be appointed as indicated in the Schedule;

5 any difference or dispute arising out of this Agreement shall be referable to arbitration in accordance with clause 3.26 of the above mentioned Architect's Appointment.

As Witness the hands of the parties the day and year first above written

Signatures:	Client	Architect
Witnesses:	Name	Name
	Address	Address
	Description	Description

20

Schedule of services and fees

Referred to in the Memorandum of Agreement dated ______

between ______
(insert name of client)

and ______
(insert name of architect or firm of architects)

for ______
(insert description of project)

Unless otherwise stated the services listed in **S1**, the conditions of appointment, and the basis of fee calculation, will be as described in the Architect's Appointment (1982 Edition ______ Revision), issued by the RIBA. Clause references relate to that document.

S1 SERVICES

Service	Clause	Fee basis (State whether percentage, time or lump sum)	Clause
Preliminary Services			
Basic Services			
Other Services			

S2 SPECIAL CONDITIONS Insert any conditions other than those in Parts 3 and 4 which are to apply to the appointment.

21

S3 **CONDITIONS NOT TO APPLY**

Insert any clauses in Parts 3 and 4 which are **not** to apply to this appointment (n.b. the alternatives in clauses 3.26 and 3.28 which are not to apply should be inserted).

S4 **PERCENTAGE FEES**

Fees based on a percentage of the total construction cost shall be calculated as follows:

S5 **LUMP SUM FEES**

Lump sum fees shall be as follows:

S6 **INTERIM PAYMENTS**

Interim payments for percentage and lump sum fees shall be
*paid monthly/quarterly/half yearly:

*paid at completion of work stages as follows:

Work stage	Proportion of fee	Cumulative total
C		
D		
E		
F G		
H J K L		

Notwithstanding these, fees in respect of work stages E, FG and HJKL shall be paid in instalments proportionate to the drawings and other work completed or the value of works certified.

*Delete whichever is inapplicable

22

S7 TIME CHARGE FEES

Rates for fees charged on a time basis shall be:

·1 for principals: £ per hour

Adjustments in the above rates shall be made at intervals of not more than 12 months on the following basis:

·2 for staff: p per £100 of gross annual income for office based staff

p per £100 of gross annual income for site based staff

S8 EXPENSES AND DISBURSEMENTS

The fees charged in accordance with **S1** and **S4** to **S7** above are inclusive of all expenses and disbursements.

or

*Expenses and disbursements shall be charged in accordance with Part 4, clauses 4.32 to 4.34.

Mileage rates shall be:

Adjustments in the above rates shall be made on the following basis:

*Delete whichever is inapplicable

S9 CONSULTANTS

The following consultants shall be appointed by the Client:

S10 SITE STAFF

The following site staff shall be appointed:

·1 by the Client

·2 by the Architect

Signed: Client Architect

Date:

23

Fig 1.3 Letter from architect to client: to enclose sketch plans and approximate estimate

Dear Sir

I have pleasure in enclosing two copies of my [*preliminary*] sketch plans for this project, prepared from the information contained in your brief. One copy is for your retention and one copy may be used to illustrate any amendments on which you would like me to advise.

You will see from the site plan dimensions that the area is rather smaller than you envisaged and this reduction in area is not helped by the positioning of the building lines laid down by the local authority. The building, therefore, is narrower than you showed on your sketches to comply with these restrictions.

With regard to the approximate cost of the building, I have prepared an estimate based on the following specification notes which amplifies the materials described in your brief as follows:

[*Here insert short description of structural frame, floor, walls and roof construction, methods of heating and lighting etc*]

On the basis of this specification and the dimensions indicated on the plan of the building, I would estimate that the cost of the project, exclusive of any professional fees or out of pocket expenses, would be in the region of £60 000, which includes the sum of £4 500 for site works including vehicular access. This estimate is based on wage and material rates applicable at this date.

When you have had an opportunity of studying the sketch proposals and considering the approximate estimate of cost, I should be pleased to meet you again to answer any queries or comments you may have on the drawings.

Yours faithfully

clearly stated. Alternative proposals should be described clearly for consideration.

Approximate estimates tend to become fixed final costs in a layman's mind. Therefore great care should be taken to ensure that the estimate is realistic and bears as close a relationship to the likely tender as possible. To do this the dimensions of the building must be fixed and a specification of structure, finish and equipment prepared so that volume or area related to standard can produce a firm realistic unit price. The usual yardstick is the cost in pounds per square metre of floor area. As will be appreciated, the storey height of the building will play a considerable part in computing the unit cost as will the complexity or otherwise of the accommodation proposed. Care must be taken to ensure that the client is fully aware of the standard of construction, finish and dimensions of the building, and that the estimate is based on these. Any variation in dimensions, specification, wage or material cost subsequent to the preparation of the estimate will vary the total.

Estimates are generally exclusive of VAT, fees and expenses and this fact should be made clear to the client. Estimated fees and expenses etc should be provided when requested.

A further meeting to clarify any matters arising from the sketch proposals is usually necessary. The suggestion contained in the letter opens the door to this and makes it clear to the client that the sketches are proposals only and that, while they are the product of much thought, they are not final and are subject to revision. Time spent at this stage to get the scheme right saves much alteration and difficulty at later stages.

Confirmation of any instructions received, either orally or by letter, should be in writing (see Fig 1.4). Where a drawing is concerned its reference should be quoted. Any amendments to the proposals should be listed in detail; clients have a habit of forgetting precise details and this could become a later embarrassment if they are not confirmed in writing.

The Memorandum of Agreement describes work stages and relates them to fee percentages which accrue at these points. Cash flow is as important to the architect as to any other business. Regular fee accounts at work stages help to keep income steady. The inclusion of out of pocket expenses in the account also assists in monitoring financial liquidity and stability.

The appointment of specialist consultants for the preparation of quantities, structural engineering, mechanical and electrical services etc is often necessary, even on small projects (see Fig 1.5). The precise time of appointment will vary with each job, but the need for any consultant is generally within the purlieu of the architect's brief under the

Fig 1.4 Letter from architect to client: acknowledgement of approval of sketch proposals

Dear Sir

I should like to confirm that, apart from the minor amendments described below, the proposals contained in my sketch drawings, number [*add drawing number or reference*], forwarded to you with my letter of [*insert date*] for your new building, are approved. The amendments to the proposals agreed at our recent meeting are as follows:

[*Insert details here*]

I should also like to confirm your instructions to proceed with the necessary action to obtain local authority approvals and competitive tenders for the works. Also I confirm your agreement to the approximate estimate of cost, amounting to £60 000.

As I indicated to you at our meeting, I have been to some expense in this matter and, in accordance with the Conditions of Engagement, fees to work stage C are now due. I therefore enclose a note of my charges to date, computed on the basis of the approximate estimate of cost, which I trust you will find in order.

Yours faithfully

Fig 1.5 Letter from architect to client: appointment of consultants

Dear Sir

With reference to our telephone conversation today, I confirm your agreement to the appointment of quantity surveyors [*consultant structural engineers, etc*] for this project. In this connection I should like to nominate Messrs [*add name and address of nominated firm*]. I have spoken to Mr [*insert name*], a partner, and his firm will be very pleased to act for you in this project.

I should be obliged if you would write direct to Messrs [*insert name*] confirming their appointment, and in the meantime I will arrange for the necessary information and drawings to be sent to them to enable them to put the work in hand.

Yours faithfully

Fig 1.6 Letter from architect to consultant: confirmation of nomination

Dear Sir

With reference to our telephone conversation, I confirm that you will be appointed to act as quantity surveyor [*structural engineer, etc*] for this project and your appointment will be made by my client [*insert name*] who will be directly responsible for any fees and charges applicable to your professional services.

Yours faithfully

Memorandum of Agreement, subject to agreement in detail by the client.

Usually the consultant is nominated by the architect and is appointed and paid by the client. Make sure that the consultant is willing to be appointed before you suggest his firm to your client (see Fig 1.6)

Most new projects incorporate specialist equipment or materials, in varying degrees. It is important to find out as soon as possible what the implications are of their selection and installation so that both the drawings and the specification can be accurately prepared to accommodate items selected.

Details of delivery period and cost are important (see Fig 1.7). Extended delivery periods for items required early in the contract may necessitate pre-contract ordering to ensure delivery when required. A long delivery period within the period of the contract is not so important for items required towards the end, such as finishing items, so long as intention to nominate or place an order through specification in the contract is implied. Care must be taken to ensure that no firm undertakings are given unless there is every likelihood of the project proceeding at any cost. Otherwise materials or equipment may be specially made for a project that is subsequently cancelled.

The provision of an approximate estimate of cost will assist in the computation of prime cost and provisional sums to be included in the contract specification or bill of quantities.

Application forms for planning approval may be obtained from the offices of the local planning authority either by post or by personal application. Copies to be submitted usually number four, and a further copy should be kept for filing and future reference. In addition it is necessary to submit certificates and notices as follows:

- Certificate A should be completed and returned when the applicant owns the freehold or is entitled to a tenancy on the property with more than ten years left unexpired.
- Certificate B should be completed when other persons who are freeholders of the land which lies within the area are affected by the planning application. The names and addresses of all such persons must be included on the certificate and to each of them must be sent a copy of Notice No. 1 to advise them of the application.

The number of drawings to be submitted will be indicated on the notes accompanying the planning application forms (see Figs 1.8 and 1.10). Care should be taken to see that the requirements of the authority are fully met in preparing the application. The block and key plans may be prepared either direct from OS maps or from extracts which can usually be obtained from the offices of the planning authority. The site forming

Fig 1.7 Letter from architect to sub-contractors: preliminary information

Dear Sirs

I have been appointed to act as architect for the above project which includes the following specialist equipment:

[*Insert here brief description of the equipment*]

[*a*] I should be pleased to receive full details and an approximate estimate of cost for the supply and installation of this equipment.

[*Or*]

[*b*] I should be pleased if you could arrange for your technical representative to make an appointment with me to discuss this equipment and its installation at an early date.

Yours faithfully

Fig 1.8 Letter from architect to local planning authority: application for approval

Dear Sir

I enclose the following in support of an application for approval under the Town and Country Planning Act 1971:

[*a*] [...] copies of Form [...] duly completed and signed. [...] copies of my drawing numbered [...] Certificate A.

[*Or*]

[*b*] [...] copies of Form [...] duly completed and signed. [...] copies of my drawing numbered [...] Certificate B.

Should there be any queries on this application I shall be pleased to discuss them with your representative.

Yours faithfully

Fig 1.9 Letter from architect to client: to enclose copies of final production drawings

Dear Sir

I have pleasure in enclosing for your use one copy each of my drawings numbered [*insert reference numbers*] being the production working drawings for this project. Could you please let me know as soon as possible whether these fully meet your requirements so that I can prepare and submit the application to the local authority for the necessary approvals as required by the Building Regulations.

Yours faithfully

the application should be coloured pink and any adjoining area under the same ownership coloured blue.

Notice of Intent forms are obtained from the local authority either by post or by personal application. Two copies are generally required and a copy should be kept for filing and future reference.

Compliance with the Building Regulations 1985 may be achieved by use of one of two alternative routes, either:

- Local Authority Building Control

or

- NHBC Building Control Services Ltd

Applicants will have to make a choice that will depend upon the acceptability of the service and the cost involved.

Using the local authority route applicants may submit full plans and elect to adopt use of the Approved Documents, appropriate standards, or well established proved methods of construction. Alternatively,

Fig 1.10 Letter from architect to local authority building control: application for approval

Dear Sir

I enclose the following in support of an application for approval under the Building Regulations 1972:

[....] copies of the Notice of Intent duly completed and signed.

[....] copies of each of my drawings numbered [....]

[*Calculation sheets and details of structural steelwork or reinforced concrete components should be included with the application and described in the letter*]

I shall be pleased to arrange attendance at your office to amend, expand or add any additional information you may require on receipt of your request.

Yours faithfully

applicants may use the Building Notice procedure (except where means of escape is involved and the building serves a designated use under the Fire Precautions Act 1971). When the Building Notice procedure is used the local authority will not pass or reject plans, as this arrangement is intended to dispense with plan examination and be devoted to site inspections. Both local authority procedures involve site inspections at various stages of the work as the final part of the control system.

Alternatively, NHBC Building Control Services Ltd is a wholly owned subsidiary of the National House Building Council and hence involved with new housing work up to four storeys in height. Applicants using this route will need to provide certain information, details of which the organisation is required to give to the appropriate local authority: this takes the form of an 'initial notice'. On acceptance of the notice by the local authority, work may start and the site inspections are carried out by NHBC Building Control Services Ltd. On completion of the works a

'final certificate' is sent to the local authority. With this system the necessary insurance cover is provided by the NHBC and fees are charged on their own scale which is different from that of the local authority prescribed fees.
The number of copies of drawings required is shown on the forms and is generally two. Care must be taken to ensure that all information is shown, otherwise the application may be rejected or extra information requested.
Only one copy of the structural details and calculations is generally required for submission, no copies being returned with the approval notices.
However carefully the drawings are prepared, building control often requires notes to be added to the drawings which, while irrelevant to the physical building process, are needed to satisfy the multitude of regulations forming the statutory requirements. It is best to state your willingness to attend to, add or amend some detail to avoid receiving a rejection on some minor matter which is often quite irrelevant to the erection of your building (see Fig 1.10). The drawings can be amended in ink and these amendments should be signed and dated.

It is not always necessary for quantities to be prepared for a building project. Price and complexity play a great part in the decision whether or not to employ a quantity surveyor. If the project requires one, the procedure for sending him information is as follows.
Two copies of all documents and drawings should be sent to the quantity surveyor to enable him to prepare his bills of quantities. These copies should all be finalised and complete with all details. It is not good practice to feed the quantity surveyor with incomplete details and information.
The draft specification can be in holograph form so long as it is legible. After the bills are prepared the specification can be corrected if required and then typed for the use of the contractor.
Agree a date with the quantity surveyor for the completion of the bills at the start of his work and keep him to the date (see Fig 1.11). Each tenderer needs two copies of the blank bill, one to be returned priced with his offer and one for his retention.

Contractors should be invited to tender for building work on the basis of the Code of Procedure for Single Stage Selective Tendering 1977 where selective competitive tenders are required. This document is produced by the National Joint Consultative Committee for Building and published by RIBA Publications Ltd. The Code assumes the use of standard forms of building contract and lays down agreed procedures for

Fig 1.11 Letter from architect to quantity surveyor: preparation of bills of quantities

Dear Sir

I enclose two copies each of the following to enable you to commence the preparation of quantities for this project:

Production working drawings numbered [...]
Schedules of windows, doors, ironmongery and manholes.
Schedule of provisional and prime cost sums.

I also enclose one copy of the draft specification on which the quantities should be based.

I confirm that tenders will be invited from four contractors on [*insert date*] and that you will supply me with sufficient copies of the bills to send two to each tenderer plus two copies for my use.

Please let me know of any queries or additional information you may require.

Yours faithfully

dealing with errors or discrepancies between the build-up of the tender and the tender figure. It is a condition of tender that obvious errors should be dealt with by the alternative selected and incorporated in the formal invitation to tender sent to each selected contractor.

Contractors are best invited to tender from a list of your own selection. Advertizing usually brings a whole shoal of fringe firms of dubious quality. The number of contractors invited to tender should be restricted to a number that will ensure not only a spirit of competitiveness but at the same time give each contractor a reasonable chance of success (see Table 1.1).

Table 1.1

Size of Contract	Max number of tenderers
Up to £50,000	5
£50,000 to £250,000	6

A preliminary enquiry should be made to selected contractors so that they can decide whether they wish to tender and to warn their estimators that the job will be coming to them for pricing if they are interested (see Fig 1.12). Certain basic information must be included in the enquiry so that the contractor is fully aware of the details of the proposed works.

Tendering documents vary depending on whether or not quantities are prepared. When the tender is prepared on the basis of drawings and specification, these must be sent out with the invitation to tender (see Fig 1.13). As they are expensive to prepare, drawings and specification should be returned with the tender for re-use during the contract. Where quantities form the basis for tender preparation, two copies of the blank bills should be sent out together with the tender form (see Fig 1.14), blank schedule of basic rates (see Fig 1.15) and label. Only the tender form, duly completed and in an envelope under the label supplied, is here returned with the offer. The label, which should be supplied to each tenderer, should – in addition to the architects' name and address – bear the job number and a code letter indicating to the architect on receipt which contractor has returned a tender. This preserves secrecy of tendering until the envelopes are opened. The period of time given to the contractor for the purpose of preparing his estimate will vary with the size and complexity of the work and the amount of measured work of a specialist nature. A minimum of four clear working weeks should be allowed with extra time allowed for work of a complicated or special nature. The client should also be informed that tenders have been invited (see Fig 1.16).

Fig 1.12 Letter from architect to prospective tenderers: preliminary invitation to tender

Dear Sirs

I have been instructed to prepare a preliminary list of contractors prepared to submit a tender for the works described in this enquiry. Your attention is drawn to details of the alternative clauses and amendments to the Standard Form of Building Contract [*Intermediate Form of Contract/Agreement for Minor Building Works*] which will be incorporated in the tender documents and which are detailed below. Tendering will be in accordance with the principles laid down in the Code of Procedure for Single-Stage Selective Tendering 1977.

Will you please advise me in writing not later than [*insert date*] whether you wish to be invited to submit a tender for these works on this basis. Inclusion in the preliminary list will in no way guarantee you will subsequently receive a formal invitation to tender. Your inability to accept this invitation will in no way prejudice future opportunities to tender for other contracts.

Yours faithfully

Schedule of Information, Alternative Clauses and Amendments to:

The Standard Form of Building Contract, Intermediate Form of Contract and Agreement for Minor Building Works.

(a) Details of works
(b) Name and address of employer
(c) Quantity surveyor [*if employed*]
(d) Structural engineer [*if employed on a supervisory basis*]

Fig 1.12 *(continued)*

(e) Location of site

(f) Approximate cost value of works

(g) Nominated sub-contractors [*if known*]

(h) Form of Contract [*insert details*]
Alternative clauses applicable [*insert details*]
Any special items

(i) Examination and correction of priced bills [*not applicable to Agreement for Minor Building Works*]
Alternative 1/Alternative 2 (1) will apply

(j) Anticipated date for possession [*insert date*]

(k) Period for completion of works [*insert period in weeks*]

(l) Date for dispatch of tender documents [*insert date*]

(m) Tender period [...] weeks

(n) Tender to be open for [...] weeks

(o) Liquidated damages at £[...] per week

Fig 1.13 Letter from architect to prospective tenderers: invitation to tender

Dear Sirs

Further to your letter of acceptance of the invitation to tender for the above, I now have pleasure in enclosing the following:

(i) Two copies of the bill of quantities.

(ii) Two copies of 1/100 drawings showing general character and arrangement of the works.

(iii) Two copies of the form of tender and schedule of rates.

(iv) [...] addressed envelope [*labels*] for the return of tender.

[*Provide alternative details of enclosures where quantities are not provided*]

(i) One copy of the specification of works.

(ii) One copy of each of drawings nos [*insert details*]

(iii) Two copies of the form of tender and schedule of rates.

(iv) Addressed envelopes [*labels*] for the return of tender.

Will you please note:

(a) The site may be inspected by arrangement with the architect [*employer*].

(b) Tendering procedure will be in accordance with the principles of the Code of Procedure for Single-Stage Selective Tendering 1977.

(c) Examination and adjustment of priced bills: Alternative 1/Alternative 2 (1) will apply.

The completed form of tender in the enclosed envelope [*label*] provided will be delivered at this office not later than 12 noon on [*insert date*].

Please acknowledge receipt of this invitation.

Yours faithfully

Fig 1.14 Example of Form of Tender

FORM OF TENDER

To Line and Wash, architects
The Muse
Olympia

Project ....
....

Client ....

Job number ....

I/We hereby agree to enter into a contract to carry out and complete the whole of the works required to be carried out in connection with the above project strictly in accordance with the drawings numbered .... specification/bill of quantities and Conditions of Contract prepared by you for the sum of £..................................... (£ . p)

I/We have included in my/our tender the provisional and prime cost sums included in the specification/bill of quantities.

The contract sum is exclusive of value added tax. The estimated liability under VAT in respect of the project works at tender stage amounts to £.....................................(£ . p) calculated at the rate of ...... %

I/We have completed and attach to this our tender a schedule of basic labour and material rates, which, should our tender be accepted, will form part of the contract for the purposes defined in the Conditions of Contract.*

I/We require the following percentages to be added to the prime cost of dayworks (as defined and published by the RICS and NFBTE)*

Fig 1.14 *(continued)*

Labour .... % Materials .... % Plant .... %*

I/We undertake to commence work on site within .... weeks of signing the Articles of Agreement and thereafter complete the works within .... weeks of commencement on site, subject to satisfactory programming and delivery dates by nominated subcontractors and suppliers, the period of the works given above to be inserted in the contract.

I/We fully understand and accept that the employer is not bound to accept the lowest or any tender and that we have visited the site of the works and are fully aware of the problems if any involved in this connection

Signature for and
on behalf of contractor ....

Name of contractor ....

Address ....

....

Date of tender ....Post code ....

(* Refers only to tenders submitted for contracts under the JCT Standard Form of Building Contract. Delete if contract form is the Agreement for Minor Building Works)

Fig 1.15 Example of Schedule of Rates

To Line and Wash, architects
The Muse
Olympia

Project ....
....

Client ....

Job number ....

SCHEDULE OF BASIC RATES

(for use with JCT Standard Form of Building Contract only)

Labour	Craftsmen	per hour
	Labourers	per hour

Material	Rate	Unit Quantity

Signature for and
on behalf of contractor ....

Name of contractor ....

Address ....
....
....

Date of tender .... Post code ....

Fig 1.16 Letter from architect to client: to advise tenders invited

Dear Sir

I have today invited tenders for the works from the following contractors who have advised me of their willingness to submit an offer:

1 Uptight Contractors Ltd.
2 Ace (Builders) Ltd.
3 J Smith and Sons Ltd.
4 A.B. Construction Co Ltd.

Tenders are to be delivered to my office not later than noon on [*insert date*] and I will immediately advise you in writing of the results.

Yours faithfully

Fig 2.1 Letter from architect to quantity surveyor: to check tenders

Dear Sir

I append a list of tenders received for this project. Will you please arrange to check the bill of quantities submitted in support of the offer by the lowest contractor and let me have your report on the arithmetical accuracy and pricing of the bill as soon as possible.

Yours faithfully

Fig 2.2 Letter from architect to all tenderers: acknowledgement of receipt of tender

Dear Sirs

I confirm the receipt of your tender for the above works which is receiving attention.

Yours faithfully

2 Placing the contract

Immediately a tender based on a bill of quantities is received the priced bill should be forwarded unopened to the quantity surveyor for an arithmetical check for accuracy and a report on the pricing of the various items (see Fig 2.1). In some cases the priced bill is not returned by the contractor who only submits the Form of Tender, waiting to hear the result of his bid before sending off his bill. In this case the quantity surveyor should be asked to obtain the priced bill from the lowest contractor.

Priced Bills of Quantities are confidential and details of pricing should not be disclosed to any person other than the architect and quantity surveyor without prior permission of the tenderer. If the quantity surveyor finds errors in the pricing he reports them to the architect, who deals with the errors in accordance with the method previously selected and referred to in the invitation to tender. These alternatives are as follows:

(a) The tenderer is given details of the errors and asked to either confirm his original offer or withdraw his tender; *or*

(b) the tenderer is given the opportunity of confirming his original offer or amending it to correct genuine errors.

If the tenderer either withdraws or on amending his tender is no longer lowest, then the next highest tender will be considered. If the tenderer stands by his tender or after amendment is still the lowest an endorsement should be added to the priced bill indicating all rates and prices (excluding preliminary items, contingencies, prime cost and provisional sums) are to be considered increased or reduced in the same proportion as the corrected total of priced items falls short or exceeds such items.

With tenders prepared on the basis of a specification and drawings providing merely a lump sum as indicated on the Form of Tender, this procedure would not apply.

The acknowledgement of the receipt of tenders should be formal and brief and should be sent to all contractors submitting a tender on the same day as tenders are due (see Fig 2.2).

Any tender received after the time and date stipulated should be returned to the contractor concerned unopened. A covering letter explaining the action taken should be attached (see Fig 2.3).

The letter from architect to client to report on tenders is a formal letter requiring formal treatment (see Fig 2.4). Unless you are in a position of social intimacy with your client it is best to keep all correspondence formal and wait for him to initiate any change.

The list of tenders received should be placed in descending order of value. Reference should be made to any estimate of cost and related to the lowest tender received. If all the tenders greatly exceed the estimate an explanation must be given for possible reasons (e.g. an increase in the size of the project after the preparation of the estimate; an increase in wage and material rates after the preparation of the estimates or an alteration in the standard of work required; an extended period between the preparation of estimates and the drawings or between the preparation of drawings and instructions to proceed to tender; lack of interest shown by contractors owing to excess of work on hand; lack of interest in the type of work proposed).

The date for the commencement of the contract is important to the client, who once the works have commenced, may well be put to inconvenience due to contractual work being carried on around him, or who may need to know precisely when he can expect to enjoy his new building. The period of the contract will have been agreed and inserted in the details included in the formal invitation to tender for the works.

In the example quoted the two highest tenders are mere formalities and not worthy of serious consideration.

The recommendation of a tender is a serious business. Personal recommendation could be disastrous if a change of direction or circumstances of a firm had materially altered its standards. If recommendation has been made by a third party you should make this plain to your client. However you will have made enquiries about the firms invited to tender before sending out the documents and only satisfactory contractors should have been included in the list. If the client insists on accepting any other tender, you will have made the right recommendation.

The reason for a client's acceptance of a tender other than the lowest submitted should always be given to the unsuccessful contractor (see Fig 2.5). He will otherwise feel justifiably aggrieved and may well suspect collusion between either the architect or client and the contractor whose tender is in fact accepted.

Confirmation of instructions to accept a tender should always indicate

Fig 2.3 Letter from architect to tenderer: return of late tender unopened

Dear Sirs

I return your tender submitted for the above works unopened.

I regret that, owing to its arrival after the time and date stipulated in the invitation to tender, your offer cannot, in this instance, be considered.

Yours faithfully

Fig 2.4 Letter from architect to client: report on tenders

Dear Sir

I have today received offers from the four contractors invited to tender for the works as follows:

1 A.B. Construction Co Ltd - £58 681.78
2 J.Smith and Sons Ltd - £60 731.00
3 Uptight Contractors Ltd - £64 332.00
4 Ace (Builders) Ltd - £68 000.00

You will see that the estimate for the works at £60 000 falls between the two lowest tenders received.

Messrs A.B. Construction Co Ltd undertake to commence work within [...] weeks after signature of the contract and complete within [...] weeks.

I have past experience of work carried out by Messrs A.B. Construction Co Ltd and have heard nothing to their detriment in recent months and I therefore recommend acceptance of their tender to you. I look forward to receiving your instructions in due course.

Yours faithfully

that this is on behalf of the client (see Fig 2.6). Otherwise, legally, the architect may find he is, in certain circumstances, under some financial liability. Always incorporate details of the accepted tender including the amount, date and name of the successful tenderer. Clients are often unbusinesslike in these as well as other matters, especially in the excitement of starting work, and reiteration of their instructions can reduce the likelihood of a mistake. This procedure is even more important when instructions are received verbally.

Agree a date either by telephone or letter for the commencement of the works with the client and agree this with the contractor before confirming in writing.

If instructions are received by telephone or during a meeting with the client, amend the start of your letter accordingly (see Fig 2.7).

Figure 2.8 is an example of the letter that can be used as a formal notification of acceptance of the tender by the client. No mention should be made or is necessary regarding any other tender.

Confirmation of the date of commencement is important as the contractor will need to start to organise his staff to deal with this new contract. A foreman will be needed and huts and equipment earmarked. Completion date is at the moment far from the thoughts of the contractor but you should mention this now to let him know that the architect and client are as interested in completion as they are in the commencement of the works.

The contract documents in most small works contracts comprise:

- A copy of the specification.
- A set of drawings identical to those from which the contractor prepared his tender.
- A contract form.

Identification of all documents is required to associate them with the tender.

The Standard Form of Contract requires the specification to be marked 'A' and this should be so marked on the front cover.

The Intermediate Form and the Agreement for Minor Building Works require no such identification.

Each specification should be endorsed on the front page as follows:

> This is the specification [*marked 'A'*]
> on pages [*insert number*]
> referred to in the contract date [*insert date*].
> [*Signature*] Employer [client]
> [*Signature*] Contractor

The inclusion of the number of pages is important to avoid any possibility of accusation of collusion.

Fig 2.5 Letter from architect to unsuccessful contractor submitting lowest tender: to advise that the tender has not been accepted by client

Dear Sir

The tender submitted by your firm for this project was the lowest received and was recommended by me for my client's acceptance. He has however accepted a higher tender amounting to [*insert sum*]

I regret that my recommendation was not followed in this matter.

Yours faithfully

Fig 2.6 Letter from architect to client: confirmation of written instructions to accept a tender

Dear Sir

Thank you for your letter of [*date*] instructing me to accept on your behalf the tender dated [*date*] amounting to [*insert amount of tender*] submitted by [*insert name of contractor*] for works to the above.

I should like to confirm that you are agreeable to work commencing on site on [*date*] and, in accordance with the period of the contract, namely [*weeks*], completion subject to the conditions of the contract should be on [*date*]. I have spoken to the contractor and he is agreeable to these dates and their insertion in the contract.

The contract documents will be prepared and forwarded to the contractor for his signature within the next few days. As soon as they are returned to me I will arrange to call on you for your signature.

Yours faithfully

Fig 2.7 Letter from architect to client: confirmation of verbal instructions to accept a tender

Dear Sir

Thank you for your instructions received verbally today [*or insert date*] etc

[*Then set out details as above*]

Fig 2.8 Letter from architect to successful contractor: acceptance of tender

Dear Sir

I have received instructions from my client to inform you that he accepts your tender dated [*date*] amounting to [*amount*] for works to the above in accordance with my drawings numbered [*insert numbers*] and specification [*or bills of quantity prepared by*].

With reference to the dates for commencement and completion of the works, my client has suggested and I confirm you have agreed that [*date*] will be suitable for the commencement of work. Consequently the date for completion calculated in accordance with the period of contract works will be [*date*].

The contract documents are being prepared and will be forwarded to you within the next few days for your signature.

Yours faithfully

The date is inserted in the contract by the architect after the signature of the contract by both employer and contractor. You will note that at this point your client becomes the 'Employer' as designated in the contract form.
The drawings could well be the set used by the contractor in the preparation of the tender. No drawings other than those on which he tendered are admissible as contract drawings.
Neither form of contract referred to specifically requires identification other than the numbers of the drawings which are inscribed in the contract. Each drawing forming part of the contract should, however, be endorsed as follows:

> This is drawing number [*insert number*] referred to in the contract dated [*insert date*].
>
> [*Signature*] Employer
> [*Signature*] Contractor

The contract form used will be that described in the specification preliminaries, bills of quantity or formal invitation to tender and no other. Certain conditions which would affect the amount of the tender would have been included and it is important that these are inscribed correctly in the contract.

2.1 The Standard Form of Contract

This requires the completion of information on pages 5, 6, 7 and 55 (Appendix). In addition certain alternative clauses need inclusion or deletion as necessary. These are Article 4 on page 7, Article 5.5 on page 8 (Arbitration), clauses 5.3.1.2 and 5.3.2 on page 14 (Master programme etc.) and clause 22 on page 23 (Insurance of the works), and the Appendix on page 55 should be completed either from the conditions included in the preliminary clauses of the specification or in standard form as shown. It is best to strike out or complete any inapplicable items rather than leave them open and likely to misinterpretation. Signature of both parties to the contract is on page 9 and signatures must be witnessed by a person neither related to nor a partner/director of the principal. When one or both of the parties executes under seal the alternatives must be used.

2.2 The Intermediate Form of Contract

Similar action needs to be taken in respect of the Intermediate Form of Contract. The information required on pages 1, 2 and 3 and the Appendix on page 23 should be completed. In addition the words 'architect' or 'supervising Officer' should be deleted throughout the document as appropriate. Certain alternative clauses should be deleted as appropriate, page 2 (Pricing by the Contractor) and Clause 6.3 (Insurance of works) as described on page 42.

2.3 The Agreement for Minor Building Works

This is a simple document to complete. The main inscriptions are on pages 1 and 2 and completion or deletions to clauses are required to Clauses 2.1, 2.3 and 2.5 on page 4 (Commencement and completion of works), Clauses 4.1, 4.2, 4.3, 4.4 and 4.5 on page 5 (Payment), Clauses 5.4 and 5.5 on page 6 (Statutory obligations) and Clauses 6.3A and 6.3B on page 6 (Insurance of works – Fire etc.). There is no Appendix to this form of agreement and signature and dating are as described for the Standard Form of Contract.

In some contracts, especially those involving maintenance works, no drawings are needed or provided. The words 'and has caused drawings numbered . . . and/or' should be deleted on page 1. Where a registered architect is in charge of the works the words 'supervising officer' should be deleted on pages 1, 2, 3, 4, 5 and 6.

By custom the contractor signs the contract documents first. These can be sent to him as he is generally fully conversant with the conditions. It is as well, however, to draw his attention to deletions and additions to the printed form and ask him to initial each one. The documents should be returned undated (see Fig 2.9).

The client (who now becomes the employer designated in the contract) is probably ignorant of the contract form and its implications to him. It is as well to ask him to call at your office to sign the contract when the opportunity can be taken to explain the main provisions of the agreement and his obligations in respect of the provision of access, instruction, insurance (if the works are alterations to existing structures), and discharge of certificates (see Fig 2.10).

The contract should be dated immediately after both parties have signed. While no stamp duty is payable on contracts or agreements under hand, a stamp duty of 50p remains payable for contracts and agreements under seal. Such contracts should be sent for stamping through the Post Office immediately they have been signed and dated.

With contracts comprising alterations to existing structures, the employer is generally required to effect the insurance of the property. Opportunity should be taken to ask him to contact his insurers without delay and certainly before the start of work on site:

Fig 2.9 Letter from architect to successful contractor: to enclose contract documents for signature and return

Dear Sir

Further to my letter to you of [*insert date*], I enclose the contract documents for your signature [*sealing*]. Will you please arrange for each drawing and the specification [*bills of quantity*] to be signed on behalf of your company in the space provided alongside the word 'contractor', and the Agreement as follows:

[*(a) Standard Form of Contract*]
Signature on page 9 [*witnessed*]; seal [*if applicable*] on page 9; initial against deletions and/or additions on pages 14, 23 and page 55.

[*Or*]

[*(b) Intermediate Form of Contract*]
Signature on page 5 [*witnessed*]; seal [*if appropriate*] on page 5; initial against deletions and/or additions on pages 2, 3, 18, 19 and 23.

[*Or*]

[*(c) Agreement for Minor Building Works*]
Signature on page 2 [*witnessed*]; initial against deletions and/or additions on pages 1, 2, 3, 4, 5, and 6.

Will you please return the documents completed within seven days, undated. I will advise you in writing when the Agreement has been signed by the employer and the date which has been ascribed.

Yours faithfully

Fig 2.10 Letter from architect to client: to seek appointment for signature of contract documents

Dear Sir

The contract documents have now been signed and returned to me by the contractor and I should be obliged if you could call at my office to complete the agreement.

Yours faithfully

Fig 2.11 Letter from architect to contractor: to advise completion of contract documents by employer

Dear Sir

The contract documents have been signed by the employer and dated [*insert date shown on contract*]. I enclose a copy of the Articles of Agreement for your retention.

I would draw your attention to the following:

date for possession [*insert date from contract*]
date for completion [*insert date from contract*].

[*a*] Two copies of the drawings and specification will be forwarded to you in the course of the next few days.

[*Or*]

[*b*] I enclose two copies each of the drawings and specification for your use.

Yours faithfully

o To inform them of the value of the contract plus a percentage equal to the professional fees chargeable to the works.
o To ask them to effect insurance to cover the above in accordance with either:
1 Clause 22C of the Standard Form of Building Contract, *or*
2 Clause 6.3 of the Intermediate Form of Contract, *or*
3 Clause 6.3B of the Agreement for Minor Building Works for the period of the contract and/or until the certificate of practical completion has been issued.

He should advise you in writing that he has done so and should produce a receipt for the premium paid if so requested by the contractor. Arrangements for access and/or entrance to the property can then be agreed so that the contractor can be instructed.

The contractor should be advised in writing when the contract documents have been signed by the employer and the date which has been inscribed (see Fig 2.11). This is normally the date on which the employer signed.

A copy of the Articles of Agreement must be sent to the contractor under Clause 5.2.1 of the Standard Form of Building Contract and Clause 1.6 of the Intermediate Form. Although this is not a contractual requirement under the Agreement for Minor Building Works, it is reasonable to deal with this in the same manner. The form should be identical to the original and completed to match precisely the conditions of tender and the original form. The copy should be inscribed:

> 'We certify this to be a true copy of the within written agreement'

and signed by two persons on behalf of the employer.

The original form, if under seal, should be deposited with the Post Office for a 50p revenue stamp to be impressed within thirty days of its execution. No stamp duty is payable for contracts executed under hand.

If the drawings, specifications and bill of quantities are ready they can be enclosed with this letter. Two copies of each are required for the contractor's use under the Standard Form and Intermediate Form; the Agreement for Minor Building Works specifies no number but two copies should be provided in the same manner.

It is advisable to get the contract documents completed before advising all and sundry of the amounts offered. Until this time the tenders are private and confidential to the employer. Builders often take a different view and tend to publicize their offers immediately after the closing date for tenders. However, as soon as the contract is settled, immediately advise the unsuccessful tenderers of the offers submitted (see Fig 2.12). Only list the value of the offers; do not give the names of individual contractors, otherwise at some future date tenderers may gain some advantage by deducing which firms may be on the tender list for future work.

Fig 2.12 Letter from architect to unsuccessful tenderers: to advise that their tender is unsuccessful

Dear Sir

With reference to your tender submitted for the above works, I regret that in this instance you were unsuccessful. Tenders were received as follows:

[*Here insert list*]

Yours faithfully

3 The contract I: general

The instigation of the initial site meeting should come from the architect who can then make his acquaintance with those contractor's representatives responsible for the contract works. Any further site meetings should, in the majority of small or medium-sized contracts, be either at regular intervals or at the instigation of the general contractor. This arrangement does not preclude the architect from calling a meeting at any time should events or circumstances warrant.
The initial site meeting will, in addition to giving architect and contractor an opportunity of meeting one another, enable the architect to set the tone and standard of the contract. Various matters will generally come onto the agenda, for example:

Layout of contractor's sheds and storage.
Setting out and location of datum level.
Position and layout of boards.
Samples of materials.
Appointment of sub-contractors.
Main services.
Access and security.

An agenda is necessary for the initial site meeting (see Fig 3.2). In addition careful notes of the names and occupations of those present, matters raised and decisions made are important and should be recorded by the architect. Any matters left unresolved for future action should also be recorded. Copies of the site meeting report should be sent to all parties within two days of the meeting to confirm in writing all decisions.
Subsequent site meetings called by the contractor should be recorded by his agent and copies of the minutes should be distributed by him. Unless these minutes are circulated quickly their effectiveness is lost.

Fig 3.1 Letter from architect to contractor: to arrange initial site meeting

Dear Sir

Will you please arrange for your agent and site foreman responsible for this contract to meet me on site at [*insert time and date*] to discuss the commencement of work.

Yours faithfully

Under the Standard Form of Building Contract the insurance of the works and, where appropriate, of the existing structure is a contractual condition under Clause 22 (see Fig 3.3):

- *Clause 22 (A)*. Where the contract is for a new building, the insurance is generally effected by the contractor in the joint names of himself and the employer. In addition he is normally required to add a percentage to the sum insured to cover fees for professional services in case of reinstatement after damage caused by one of the risks enumerated. The policy must be with an insurer approved by the architect and the policy and premium receipts deposited with him for safe keeping.
- *Clause 22 (B)*. This clause gives the option at tendering stage for the employer to bear responsibility for insurance of new building contracts. He must maintain a proper insurance policy and the receipt can be demanded for inspection by the contractor.
- *Clause 22 (C)*. This clause refers to the insurance, for which the employer is responsible, of a building, alteration works and unfixed materials. The employer must maintain a proper insurance policy and the receipt can be demanded for inspection by the contractor.

Under the IFC 84 the insurance of the works and/or existing structure is a contractural condition under Clause 6.3.

- *Clause 6.3A*. Where the contract is for a new building, the contractor is required to insure in the joint names of both employer and contractor against the perils defined in Clause 8.3 plus (if so required in the contract Appendix) a percentage to cover professional fees.

Fig 3.2 Report: site meeting

New Building
South Road
New Town
For Messrs Blank & Co Ltd.
Job Number 259

Initial meeting held on site on [*insert time and date*].

Present: I. Patcham, site manager, A.B. Construction Co Ltd.
W. Smith, foreman
I. Jones, architect

1 Temporary sleeper access to site agreed in position of final access road. Building lines laid down for agreement with building control.

2 Datum level on existing manhole confirmed for transference to site.

3 Position of site huts and material storage agreed.

4 Samples of materials as specified requested for:
(i) facing bricks;
(ii) common bricks for foundations and manholes;
(iii) building sand;
(iv) ballast of all grades.
Samples to be on site for inspection within seven days.

5 Work on site to commence [*insert date agreed*].

[*insert date*] Signed: I. Jones RIBA, architect

Copies to: Contractor (2)
Employer
Architect
Quantity surveyor
Consultant engineer
Clerk of works etc

Messrs Line & Wash, architects
The Muse
Olympia
London

Fig 3.3 Letter from architect to employer: insurance of works

A STANDARD FORM OF BUILDING CONTRACT

Dear Sir

Under Clause

[22B *if the insurance of a new building is to be borne by the employer*]

[*Or*]

[22C *if the insurance of alterations or extensions to a building is to be borne by the employer*]

B INTERMEDIATE FORM OF CONTRACT

Dear Sir

Under Clause

[6.3B *if insurance of new building is to be borne by the employer*]

[*Or*]

[6.3C *if insurance of alteration works is to be borne by employer*]

of the contract dated ... between Messrs [*insert names*] and yourself, you are required to effect insurance of the works. I would be obliged if you would kindly arrange for cover in accordance with the terms of this contract clause, a copy of which I enclose for your use, and advise me in due course when this cover has been effected and with which insurers.

C AGREEMENT FOR MINOR BUILDING WORKS

Dear Sir

Under Clause 6.3B of the contract dated ... between Messrs [*insert name*] and yourself, you are required to effect insurance of the works, existing structures together with the contents, and all unfixed materials and goods.

I should be obliged if you would kindly arrange for cover in accordance with the terms of this contract clause, a copy of which I enclose for your use, and advise me when the cover has been effected and with which insurers.

Yours faithfully

○ *Clause 6.3B.* Where the contract is for a new building, the employer is required to insure against the risks as defined above.

○ *Clause 6.3C.* Where the contract is for works to existing structures, the employer is required to insure against the risks as defined above.

Under the Agreement for Minor Building Works the insurance of the works and, where appropriate of the existing structure, is a contractual condition under Clause 6.

The insurance for all these contingencies must be at least for the value of the contract plus any percentage required for professional fees for reinstatement in the event of damage by one of the risks enumerated in the particular clause of the contract form employed (see Fig 3.4).

The architect must satisfy himself that all the conditions are fully met and should preferably advise the other party to the contract when he is satisfied that proper insurance has been effected and the premium/s paid.

Fig 3.4 Letter from architect to contractor: insurance of works

Dear Sir

Under Clause 22A of the Standard Form of Building Contract [*Clause 6.3A of the IFC / Clause 6.3A of the Agreement for Minor Building Works*] you are required to insure the works in the joint names of the employer and contractor for the risks enumerated for the full value of the contract [*plus ... per cent as included in the Appendix to the Conditions to cover professional fees*]. The value is to include all works executed together with all unfixed materials and goods as defined in the clause.

Will you please advise me of your proposed insurers for my approval and as soon as possible forward to me details of your policy and copy of current premium receipt for retention.

Yours faithfully

Instructions issued by the architect must be in writing. The Standard Form of Building Contract (Clause 4.3) requires confirmation of any oral instruction issued by the architect; this may be by the contractor who must confirm the instruction to the architect within seven days. If the architect does not dissent from the terms and description of the instruction contained in the contractor's confirmation within a further seven days, this will then take effect subject to the provision contained in Clause 4.3.2.

The contractor may, under Clause 4.2, challenge the architect's right to issue an instruction. If he does not comply with the instruction within seven days of receipt of written notice the employer may employ and pay another to carry out the work and deduct the cost from any monies due to the contractor.

The standard printed Architect's Instruction sheet issued by the RIBA (see p.47) is not a variation order although in many cases it may be so. It is a general purpose form on which, for administrative tidiness, many matters which relate to the running of the contract may be included, for example:

> Placing of nominated sub-contractors and orders for suppliers' materials and equipment.
> Approval of samples.
> Agreement to the appointment of sub-contractors.
> Notification of unacceptable work or materials.
> Variation in scope or extent of works.

The heading to the sheet must be completed and the date of the contract included. If the instruction is issued by the architect before the contract is dated, the word 'dated' should be erased and the words 'not yet dated' inserted. Each instruction should bear a consecutive number and each item should be similarly identified. Always date and sign the instruction. It is not necessary to cost the instruction, but this could be considered advisable if there is any likelihood of expenditure in excess of that allowed. If a quantity surveyor is retained an extra copy should be sent to him for pricing and return as soon as possible. Two copies should be sent to the contractor as required under the contract, and a copy to the employer for his records.

The IFC (Clause 3.5.1) also requires all instructions issued by the architect to be in writing and if within seven days of receipt of such written instructions the contractor has not complied the employer may make other arrangements for the work described and recover the cost from the contractor. Under Clause 3.5.2 the contractor may challenge

Architect:
address:

Employer:
address:

Contractor:
address:

Works:
situated at:

Architect's instruction

Job reference:
Serial No:
Issue date:

Original to Contractor

Under the terms of the Contract dated

I/We issue the following instructions. Where applicable the Contract Sum will be adjusted in accordance with the terms of the relevant Condition.

Instructions	Office use: Approx costs £ omit	£ add

SPECIMEN

Signed Architect

Amount of Contract Sum	£
± Approximate value of previous instructions	£
	£
± Approximate value of this instruction	£
Approximate adjusted total	£

Copies to: Employer, Quantity Surveyor, Clerk of works, Site, Structural consultant, Services consultant, Electrical consultant, Nominated Sub-Contractors, File

Architect:
address:

Employer:
address:

Contractor:
address:

Works:
situated at:

Clerk of works direction

Job reference:
Serial No:
Issue date:

Original to Contractor

Under the terms of the Contract dated ______________________

I issue this day the following direction.

This direction shall be of no effect unless confirmed in writing by the Architect within 2 working days and does not authorize any extra payment.

Direction	Architect's use
	Covered by AI No:
Signed ______________________ Clerk of works	

Copies to: __ Architect __ File

the architect's right to issue an instruction and the architect must comply with the request.

The Agreement for Minor Building Works requires that oral instructions must be confirmed in writing by the architect (Clause 3.5) within two days.

The standard RIBA Architect's Instruction sheet may be used for any such purpose without deletion or amendment.
The clerk of works, appointed by the employer under the direction of the architect, may give direction to the contractor on matters on which the architect is empowered to issue instructions.
To do so the clerk of works completes his printed direction in triplicate, issuing the original to the contractor's site representative, one copy direct to the architect, and retaining a further copy for his use.
The Clerk of Works' Direction (see p.48) must, under Clause 12 of the Standard Form of Building Contract, be confirmed in writing by the architect within two working days. This is done by the issue of an Architect's Instruction referring specifically to the Clerk of Works' Direction and serial number.

The IFC (Clause 3.10) gives the employer authority to appoint a clerk of works to act solely as inspector under the directions of the architect but makes no other direction as to his duties or responsibilities.
Neither the IFC or the Agreement for Minor Building Works make any direction in respect of Clerk of Works' Directions.

Approval to the sub-letting of any portion of the works must be obtained from the architect before sub-contracts are placed. At the initial site meeting the contractor should be asked for a list of any sub-contractors he wishes to employ, and this list should be sent to the architect in time for a check to be carried out on the capabilities of the sub-contractor.
Approval or rejection may be made either by Architect's Instruction or by letter (see Fig 3.5). In any event it should be in writing. Sub-letting is a fact of life in the building industry, and approval should not lightly be withheld.
Approval to assignment or sub-letting is required under Clause 19 of the Standard Form of Building Contract, Clause 3.2 of the Intermediate Form and Clause 3.2 of the Agreement for Minor Building Works.

The Standard Form of Building Contract (Clause 7) lays responsibility on the architect to determine the levels required for the works and also to supply the proper dimensions for the works.

Fig 3.5 Letter from architect to contractor: sub-letting

Dear Sir

With reference to your request to approve the sub-letting of works under this contract I have no objection to the employment of the following sub-contractors for the work described:

Messrs R. Brown & Co Ltd, earthworks
Messrs I. Jones Ltd, brick and blockwork
Messrs T. Spread, plastering and screeds

Yours faithfully

Fig 3.6 Letter from contractor to architect: datum levels and setting out of works

Dear Sir

Will you please confirm the datum level for the works. Your drawing W/1/259 indicates the datum to be the existing manhole cover situated in the south-west corner of the site. This manhole does not now exist. We presume it was removed during site clearance.

We have checked the figured dimensions on the drawings and find a discrepancy of 112 mm on the total length of the south elevations. Will you please advise us immediately of where the discrepancy occurs as our foreman will commence setting out the works on Thursday of this week.

Yours faithfully

Often datum marks are moved or destroyed between the original survey and commencement of work. The contractor will then require the architect to re-establish a datum for his use (see Fig 3.6). If no datum has been indicated on the drawings the contractor is within his rights to obtain the architect's approval to his datum before proceeding with the work. Temporary datum points should be properly established, either as fixed solid components such as a manhole cover in the road, or with a timber peg, securely concreted in to prevent movement.

Discrepancies on figured dimensions must be investigated immediately and any errors rectified in writing (see Fig 3.7). Delay can cause the contractor loss and he could well claim extra payment or an extension of time.

Any divergence between any two of the documents should be reported to the architect by the contractor in writing. The architect must issue such instructions as are necessary to clarify the matter. The Standard Form of Building Contract (Clause 2.3) describes and enumerates the docu-

Fig 3.7 Letter from architect to contractor: datum levels and discrepancy of dimensions

Dear Sir

With reference to the datum level for this job, the level indicated on the manhole was transferred from the bench mark existing on the corner buttress of the chapel further down the road. This bench mark value 101.25 has now been transferred to a new datum on site for your use.

We have checked the figured dimensions and agree a discrepancy of 112 mm. This is due to an error in the dimension of the pier by the main entrance which should be 338 mm and not 225 mm as shown.

Your foreman on site has been advised of these matters and his drawings amended.

Yours faithfully

ments, between any two of which divergence must be clarified. These documents are as follows:

(a) the contract drawings
(b) the specification
(c) any Instruction issued by the architect under the conditions of contract
(d) any drawings or documents issued by the architect under Clause 5.3.1.1, 5.4 or 7.

The Intermediate Form of Contract (Clause 1.4) makes no reference to levels or dimensions but specifically requires the architect to issue instructions to correct inconsistencies or errors. Clause 3.9 deals with the contractual requirements in this respect and how inaccuracies are to be corrected.

The Agreement for Minor Building Works makes no reference either to datum levels or to setting out.

In all contracts, the architect should check the setting out of the works at ground level to ensure that the building is correctly dimensioned and square.

Although a letter has been used to confirm and resolve the discrepancy in this instance (see Fig 3.8), the matter could well have been dealt with solely by the issue of an Architect's Instruction. In any event a letter such as the example illustrated must have copies sent to all relevant parties, for example:

- The quantity surveyor (if appointed).
- The consultant engineer or any other consultant (if he is involved).
- The clerk of works (if employed).

The resolution of discrepancies between contract drawings and/or bills of quantities/specification is required under Clause 2 of the Standard Form of Building Contract and Clause 1.4 of the Intermediate Form of Contract. Under the Agreement for Minor Building Works inconsistencies shall be corrected under Clause 4.1.

Under Clause 34 of the Standard Form of Building Contract all antiquities discovered on the site are the property of the employer. In addition, in many historic towns it is a duty to inform the local authority of the discovery of antiquities so that these may be properly excavated and investigated by suitably qualified experts.

This procedure can involve the contractor in loss owing to delay in proceeding with the works, and he is entitled to due compensation for

Fig 3.8 Letter from architect to contractor: discrepancy between contract drawings and specification

Dear Sir

Our attention was drawn on site today to a discrepancy in concrete mixes for foundations between the drawings and the specification. The description 'concrete mix A' shown on the drawings is incorrect; that included in the specification 'concrete for all foundations to be mix B' is correct and should be used throughout.

We enclose our official instruction to resolve this discrepancy.

Yours faithfully

Fig 3.9 Architect's Instruction, from architect to contractor: discovery of antiquities

[27] The additional sum of £150 is hereby agreed and authorized to be added to the contract sum as a direct loss to the contractor as a result of delays to the construction of the foundations between [*insert dates*] during the excavation of an Iron Age burial pit under the boiler room. Revisions to the foundations, additional excavation and filling will be measured separately (or are agreed as dayworks).

this loss and/or an extension of time granted. In addition variations to the substructure, excavations and filling may occur, and directions as to the methods of valuation may be given.
The Architect's Instruction is usually used to deal with this matter (see Fig 3.9).

No direction is given under the IFC and Agreement for Minor Building Works.

The Standard Form of Tender for Nominated Sub-Contractors NSC/1 (see pp. 55–66) is for use with the Standard Form of Building Contract only. It is a general purpose document enabling both preliminary enquiries and final tenders to be obtained by insertion of relevant data.
Page 1 comprises the tender form, the headings to be completed by the architect and the alternatives of sub-contract or tender correctly deleted. The remainder, comprising the offer and daywork percentages as appropriate, are completed and the form signed by the tenderer.
Page 2 covers various stipulations with regard to the offer and in Clause 4 the subcontractor must insert the acceptance period.
Pages 3 to 6 inclusive form Schedule 1 incorporating particulars of the main contract and sub-contract. The heading and Clauses 1 to 14 are completed by the architect and Appendix A and B on pages 7 and 8 as appropriate by the sub-contractor and signed by him.
Pages 9 to 12 inclusive comprise Schedule 2, being the particular conditions relating to the sub-contract offer, and are completed in part by the sub-contractor and the remainder by the contractor in agreement with the proposed sub-contractor on receipt of the architect's preliminary notice of nomination under Clause 35.7.1 of the Standard Form of Building Contract.
All nominated sub-contractors are required to sign a JCT Standard Form of Employer/Nominated Sub-contractor Agreement. This form is issued in two variations:

- *Agreement NSC/2* is used where sub-contract work is nominated in accordance with clauses 35.6 to 35.10 (ie., where the Tender Form NSC1 is used to obtain a sub-contract quotation) (see pp. 67–70).
- *Agreement NSC/2a* is used where sub-contract work is nominated in accordance with clauses 35.11 and 35.12 (ie., where NSC1 is not used to obtain a sub-contract quotation) (see pp. 80–83).

The contractor may object to the nomination of a sub-contractor but he must do so either not later than the return to the architect of the Tender NSC/1, if this is used, or not later than seven days after receipt of an

Tender NSC/1

JCT

JCT Standard Form of Nominated Sub-Contract Tender and Agreement

See "Notes on the Completion of Tender NSC/1" on page 2.

Main Contract Works:[a]

Job reference:

Location:

Sub-Contract Works:

To: The Employer and Main Contractor [a]

We______________________________________

of ______________________________________

______________________________Tel. No: ____________

offer, *upon and subject to the stipulations overleaf,* to carry out and complete, as a Nominated Sub-Contractor and as part of the Main Contract Works referred to above, the Sub-Contract Works identified above in accordance with *the drawings/specifications/bills of quantities/schedule of rates for the Sub-Contract Works which are annexed hereto, numbered

and signed by ourselves and by the Architect/Supervising Officer; and the Particular Conditions set out in Schedule 2 when agreed with the Main Contractor; and JCT Sub-Contract NSC/4 which incorporates the particulars of the Main Contract set out in Schedule 1.

*for the VAT-exclusive Sub-Contract Sum of £____________________________

______________________________________(words)

or *for the VAT-exclusive Tender Sum of [b]£____________________________

______________________________________(words)

The daywork percentages (Sub-Contract NSC/4 clause 16·3·4 or clause 17·4·3) are:

Definition[c]	Labour %	Materials %	Plant %
RICS/NFBTE			
RICS/ECA			
RICS/ECA (Scotland)			
RICS/HVCA			

The Sub-Contract Sum/Tender Sum and percentages take into account the 2½% cash discount allowable to the Main Contractor under Sub-Contract NSC/4.

Signed by or on behalf of the Sub-Contractor

Date

Approved by the Architect/Supervising Officer on behalf of the Employer

Date

ACCEPTED by or on behalf of the Main Contractor subject to a nomination instruction on Nomination NSC/3 under clause 35·10 of the Main Contract Conditions

Date

*Delete as applicable

Page 1

Tender NSC/1

Stipulations

1. Only when this Tender is signed on Page 1 on behalf of the Employer as 'approved' and the Employer has signed or sealed (as applicable) the Agreement NSC/2 do we agree to be bound by that Agreement as signed by or sealed by or on behalf of ourselves.

2. If the identity of the Main Contractor is not known at the date of our signature on page 1 we reserve the right within 14 days of written notification by the Employer of such identity to withdraw this Tender and the Agreement NSC/2 notwithstanding any approval of this Tender by signature on page 1 on behalf of the Employer.

3. We reserve the right to withdraw this Tender if we are unable to agree with the Main Contractor on the terms of Schedule 2 of this Tender (Particular Conditions).

4. Without prejudice to the reservations in 2 and 3 above this Tender is withdrawn if the nomination instruction (Nomination NSC/3) is not issued by the Architect/Supervising Officer (with a copy to ourselves) under the Main Contract Conditions clause 35·10 within [d] of the date of our signature on page 1 or such other, later, date as may be notified in writing by ourselves to the Architect/Supervising Officer.

5. Any withdrawal under 2, 3 and 4 above shall be at no charge to the Employer except for any amounts that may be due under Agreement NSC/2.

Notes

[a] For names addresses and telephone numbers of Employer, Main Contractor, Architect/Supervising Officer and Quantity Surveyor see Schedule 1.

[b] Alternative for use where the Sub-Contract Works are to be completely re-measured and valued.

[c] If more than one Definition will be relevant set out percentage additions applicable to each such Definition. There are four Definitions which may be identified here; those agreed between the Royal Institution of Chartered Surveyors and the National Federation of Building Trades Employers; the Royal Institution and the Electrical Contractors Association; the Royal Institution and the Electrical Contractors Association of Scotland and the Royal Institution and the Heating and Ventilating Contractors Association.

[d] Sub-Contractor to insert acceptance period.

[e] See Schedule 2, Item 11.

[f] Delete the two alternatives not applicable. Insert 35·7 or 36·8 percentage.

[g] Insert the same description as in the Main Contract Articles of Agreement.

[h] Delete editions/Supplement as applicable. Insert date of revision.

[i] Insert office address as appropriate.

[j] If a later Completion Date has been fixed under clause 25 this should also be stated here.

[k] Standard Form Local Authorities WITH Quantities edition only.

[l] This information, unless included in the Sub-Contract Specification or Bills of Quantities should be given by repeating here or attaching a copy of the relevant section of the Preliminaries Bill of the Main Contract Bills (or of the main Contract Specification).

[m] Sub-Contractor to complete this Appendix and attach further sheet if necessary.

[n] Date to be inserted by the Architect/Supervising Officer.

[o] To be deleted by the Architect if fuels not to be included. See NSC/4 clause 35·2·1 and clause 36·3·1.

[p] Only applicable where the Main Contract is let on the Standard Form of Building Contract, Local Authorities Edition, WITH Quantities.

[q] If both specialist engineering formulae apply to the Sub-Contract the percentages for use with each formula should be inserted and clearly identified.

[r] The weightings for sprinkler installations may be inserted where different weightings are required.

[s] To be completed by the Architect/Supervising Officer.

[t] The Sub-Contractor will set out here (or on an attached sheet if necessary) details of the carrying out of the Sub-Contract Works as a preliminary indication to the Architect and Contractor. In regard to periods indicate when each is to start. Adaptation will be needed where a phased completion is required.

[u] Not including period required by Architect for approval.

[v] The Contractor to complete in agreement with the Sub-Contractor details of the programme for carrying out the Sub-Contract Works (which must include the subjects set out in 1A and 1B) and to be inserted here or on an attached sheet initialled by the Contractor and Sub-Contractor. The details at 1A and 1B (and in any sheet attached thereto) must then be deleted and the deletion initialled by the Contractor and the Sub-Contractor.

[w] Attention is called to the Order of Works, if any, stated in Schedule 1, Item 11.

[x] The Sub-Contractor will set out here or on an attached sheet as a preliminary indication to the Architect and Contractor details of the attendances which he requires under the headings (a) to (g) which have been extracted from SMM, 6th Edn., B.9.3. These attendances to be supplied at no cost to the Sub-Contractor.

[y] The Contractor to complete and set out in agreement with the Sub-Contractor any alterations to any of the detail of the attendances set out under the headings (a) to (g) in 3·A.

[z] The Contractor to complete in agreement with the Sub-Contractor. This item must include any limits of indemnity which are required in respect of insurances to be taken out by the Sub-Contractor. See NSC/4, clause 7.

[aa] Any special conditions or agreements affecting the employment of labour which the Sub-Contractor wishes to raise should be inserted here.

[bb] The Contractor to complete in agreement with the Sub-Contractor. The details of 5·A. must then be deleted and the deletion initialled by the parties.

[cc] See Sub-Contract NSC/4, clause 24. The Contractor to complete by inserting the names and addresses after agreement with the Sub-Contractor.

[dd] The Contractor to ensure completion as appropriate after agreement with the Sub-Contractor. See Note [l] in Sub-Contract NSC/4, clause 20A on "self-vouchering".

[ee] The Contractor to delete in agreement with the Sub-Contractor.

[ff] The Sub-Contractor will set out here or on an attached sheet any matters he wishes to agree with the Contractor.

[gg] The Contractor to complete in agreement with the Sub-Contractor. The details of 9·A. must then be deleted and the deletion initialled by the parties.

Page 2

Tender NSC/1 Schedule 1

Schedule 1: Particulars of Main Contract and Sub-Contract

Names and addresses of:

Employer: Tel. No:

†Architect/Supervising Officer: Tel. No:

Quantity Surveyor: Tel. No:

Main Contractor: Tel. No:

1. Sub-Contract Conditions: Sub-Contract NSC/4, appropriate to the Standard Form of Building Contract edition identified in item 5 of this Schedule, unamended: 19 edition (revised..............................). To be executed forthwith after Architect/Supervising Officer's nomination on Nomination NSC/3.[e]

2. Sub-Contract Fluctuations: NSC/4[f] Clause 35 (see also Appendix A)
or Clause 36 (see also Appendix A)
35·7 or 36·8......................%
or Clause 37 (see also Appendix B)

3. Main Contract Appendix and entries therein: (see item 10 pages 4 and 5) Where relevant will apply to the Sub-Contract unless otherwise specifically stated here. The entry relating to clause 37 of the Main Contract Conditions is for information of the Sub-Contractor only.

4. Main Contract Works:[g]

5. Form of Main Contract Conditions: Standard Form of Building Contract, 1980 edition.

Local Authorities/Private edition/WITH/WITH APPROXIMATE/WITHOUT Quantities (revised............................)[h]

Sectional Completion Supplement

6. Inspection of Main Contract: The unpriced *Bills of Quantities/Bills of Approximate Quantities/Specification (which incorporate the general conditions and preliminaries of the Main Contract) and the Contract Drawings may be inspected by appointment at: [i]

*Delete as applicable

†Note: The expression 'Supervising Officer' is applicable where the nomination instruction will be issued under the Local Authorities Edition of the Standard Form of Building Contract and by a person who is not entitled to the use of the name 'Architect' under and in accordance with the Architects (Registration) Acts 1931 to 1969. If so, the expression 'Architect' shall be deemed to have been deleted throughout this Tender including the Schedules and in Agreement NSC/2. Where the person who will issue the nomination instruction is entitled to the use of the name 'Architect' the expression 'Supervising Officer' shall be deemed to have been deleted.

Page 3

Tender NSC/1 **Schedule 1**

7. Execution of Main Contract:	*is/is to be *under hand/under seal
8. Main Contract Conditions – alternative etc. provisions:	Architect/Supervising Officer: *Article 3A/Article 3B WITHOUT Quantities editions only: *Article 4A/Article 4B master programme: Clause 5·3·1·2 *deleted/not deleted Works insurance: *Clause 22A 22B 22C insurance: Clause 21·2·1 Provisional sum *included/not included

9. Main Contract Conditions – any changes from printed Standard Form identified in item 5:

10. Main Contract: Appendix and entries therein

	Clause etc	
Statutory tax deduction scheme – Finance (No. 2) Act 1975	Fourth recital and 31	Employer at Date of Tender *is a 'contractor'/is not a 'contractor' for the purposes of the Act and the Regulations
Settlement of disputes – Arbitration	Article 5·1	Articles 5·1·4 and 5·1·5 apply (see Article 5·1·6)
Date for Completion	1·3 [i]	________
Defects Liability Period (if none other stated is 6 months from the day named in the Certificate of Practical Completion of the Works).	17·2	________
Insurance cover for any one occurrence or series of occurrences arising out of one event	21·1·1	£ ________
Percentage to cover professional fees	22A	________
Date of Possession	23·1	________
Liquidated and Ascertained Damages	24·2	at the rate of £ ________ per ________

Page 4

*Delete as applicable

Tender NSC/1 Schedule 1

10. Main Contract: Appendix *continued*

	Clause	
Period of delay:	28·1·3	
(i) by reason of loss or damage caused by any one of the Clause 22 Perils	28·1·3·2	________
(ii) for any other reason	28·1·3·1, 28·1·3·3 to ·3·7	________
Period of Interim Certificates (if none stated is one month)	30·1·3	________
Retention Percentage (if less than five per cent)	30·4·1·1	________
Period of Final Measurement and Valuation (if none stated is 6 months from the day named in the Certificate of Practical Completion of the Works)	30·6·1·2	________
Period for issue of Final Certificate (if none stated is 3 months)	30·8	________
Work reserved for Nominated Sub-Contractors for which the Contractor desires to tender	35·2	________
†Fluctuations	37	*Clause 38/Clause 39/Clause 40
Percentage addition	*38·7/39·8	________
†Formula Rules	40·1·1·1	
	rule 3	Base Month ________ 19___
	rule 3	Non-Adjustable Element [k] ________% (not to exceed 10%)
	rule 10	Part I/Part II of Section 2 of the Formula Rules is to apply.

†Note: Clause 40 and Formula Rules entries are not applicable where in item 5 it is stated that the WITHOUT Quantities Conditions apply.

11. Order of Works: Employer's requirements affecting the order of the Main Contract Works (if any).

*Delete as applicable

Tender NSC/1 **Schedule 1**

12. Location and type of access:

13. Obligations or restrictions imposed by the Employer not covered by Main Contract Conditions:
(e.g. in Preliminaries in the Contract Bills)[I]

14. Other relevant information: (if any)

Signed by or on behalf of the Sub-Contractor Information noted ________________

Page 6

Tender NSC/1 Schedule 1[m] Appendix (A)

Fluctuations

See Tender NSC/1 page 3, item 2

Notes:
Complete column (1) where Sub-Contract NSC/4, clause 35 applies.
Complete columns (1)-(4) where Sub-Contract NSC/4, clause 36 etc. applies.

The 'Date of Tender'[n] for the purposes of Sub-Contract NSC/4, clauses 35/36.

is ______________________________

(1) Materials, goods, electricity and fuels[o]	(2) Rate (or appropriate standard price list description)	(3) Discounts	(4) Unit to which rate applies

Signed by or on behalf of the Sub-Contractor

Date

Page 7

Tender NSC/1

Schedule 1 Appendix (B)

Fluctuations

See Tender NSC/1 page 3, item 2

Note: Architect and Sub-Contractor to complete this Appendix as appropriate where Sub-Contract NSC/4, clause 37 applies.

1. 37·1 – Nominated Sub-Contract Formula Rules are those dated ________________ 198__
   *Part I/Part III of these Rules applies
2. 37·3·3 and ·3·4 – Non-Adjustable Element[p] ________________ % (not to exceed 10%)

3. 37·4 – List of Market Prices

4. Nominated Sub-Contract Formula Rules
   rule 3 (Definition of Balance of Adjustable Work)
   Any measured work not allocated to a Work Category

   ____

   ____

   Base Month (rule 3) ________________

   Date of Tender (rule 3) ________________

   rule 8 Method of dealing with 'Fix-only' work

   rule 11(a) Part I only: the Work Categories applicable to the Sub-Contract Works

   rule 43 Part III only: Weightings of labour and materials – Electrical Installations or Heating, Ventilating and Air Conditioning Installations[q]

	Labour	Materials
Electrical	______%	______%
Heating, Ventilating and Air Conditioning[r]	______%	______%
	______%	______%

rule 61a Adjustment shall be effected
*upon completion of manufacture of all fabricated components
*upon delivery to site of all fabricated components

rule 64 Part III only: Structural Steelwork Installations:
(i) Average price per tonne of steel delivered to fabricator's work
£ ________________

(ii) Average price per tonne for erection of steelwork
£ ________________

rule 70a Catering Equipment Installations:
apportionment of the value of each item between
(i) Materials and shop fabrication £ ________________

(ii) supply of factor items £ ________________

(iii) site installations £ ________________

Signed by or on behalf of the Sub-Contractor

Date

Page 8

*Delete as applicable

Tender NSC/1 Schedule 2

Schedule 2: Particular Conditions

Note: When the Contractor receives Tender NSC/1 together with the Architect/Supervising Officer's preliminary notice of nomination under clause 35·7·1 of the Main Contract Conditions then the Contractor has to settle and complete any of the particular conditions which remain to be completed in this Schedule in agreement with the proposed Sub-Contractor. The completed Schedule should take account not only of the preliminary indications of the Sub-Contractor stated therein, but also of any particular conditions or requirements of the Contractor which he may wish to raise with the Sub-Contractor.

1.A Any stipulation as to the period/periods when Sub-Contract Works can be carried out on site:[s]

to be between ______________________ and ______________________

Period required by Architect to approve drawings after submission ______________________

1.B Preliminary programme details[t] (having regard to the information provided in the invitation to tender)

Periods required:

(1) for submission of all further sub-contractors drawings etc. *(co-ordination, installation, shop or builders' work, or other as appropriate)*[u] ______________________

____

____

(2) for execution of Sub-Contract Works: off-site ______________________

on-site ______________________

Notice required to commence work on site ______________________

1.C Agreed programme details (including sub-contract completion date: see also Sub-Contract NSC/4, clause 11·1)[v]

2. Order of Works to follow the requirements, if any, stated in Schedule 1, item 11[w]

Page 9

Tender NSC/1 **Schedule 2**

3.A Attendance proposals (other than †general attendance).[x]

(a) Special scaffolding or scaffolding additional to the Contractor's standing scaffolding.

(b) The provision of temporary access roads and hardstandings in connection with structural steelwork, precast concrete components, piling, heavy items of plant and the like.

(c) Unloading, distributing, hoisting and placing in position giving in the case of significant items the weight and/or size. (To be at the risk of the Sub-Contractor).

(d) The provision of covered storage and accommodation including lighting and power thereto.

(e) Power supplies giving the maximum load.

(f) Maintenance of specific temperature or humidity levels.

(g) Any other attendance not included under (a) to (f) or as †general attendance under Sub-Contract NSC/4, paragraph 27·1·1.

†Note: For general attendance see clause 27·1·1 of Sub-Contract NSC/4 which states: "General attendance shall be provided by the Contractor free of charge to the Sub-Contractor and shall be deemed to include only use of the Contractor's temporary roads, pavings and paths, standing scaffolding, standing power operated hoisting plant, the provision of temporary lighting and water supplies, clearing away rubbish, provision of space for the Sub-Contractor's own offices and for the storage of his plant and materials and the use of messrooms, sanitary accommodation and welfare materials." See SMM, 6 edn., B.9.2.

Page 10

Tender NSC/1 Schedule 2

3.B .[y]

4. Insurance[z]

5.A Employment of Labour – Special Conditions or Agreements[aa]

5.B Employment of Labour – Special Conditions or Agreements[bb]

Page 11

Tender NSC/1 **Schedule 2**

6. The Adjudicator is:[cc]

The Trustee – Stakeholder is:[cc]

7. Finance (No. 2 Act 1975 – Statutory Tax Deduction Scheme[dd]

1. The Contractor* is/is not entitled to be paid by the Employer without the statutory deduction referred to in the above Act or such other deduction as may be in force;

2. The Sub-Contractor* is/is not entitled to be paid by the Contractor without the above-mentioned statutory deduction or such other deduction;

3. The evidence to be produced to the Contractor for the verification of the Sub-Contractor's tax certificate (expiry date 19) will be:

8. Value Added Tax – Sub-Contract NSC/4
Clause 19A/19B[ee] (alternative VAT provisions) will apply.

9.A Any other matters (including any limitation on working hours).[ff]

9.B [gg]

10. Any matters agreed by the Architect/Supervising Officer and Sub-Contractor before preliminary notice of nomination.[s]

11. Sub-Contract NSC/4 Edition as identified in Schedule 1, Item 1 to be executed under hand/under seal[ee] forthwith after Architect/Supervising Officer's nomination on Nomination NSC/3.

The above Particular Conditions are agreed

Signed by or on behalf of the Sub-Contractor — Date

Signed by or on behalf of the Main Contractor — Date

Page 12 *Delete as applicable

Agreement NSC/2

JCT

JCT Standard Form of Employer/Nominated Sub-Contractor Agreement

Agreement between a Sub-Contractor to be nominated for Sub-Contract Work in accordance with clauses 35·6 to 35·10 of the Standard Form of Building Contract for a main contract and the Employer referred to in the main contract.

Main Contract Works:

Location:

Sub-Contract Works:

This Agreement

The date to be inserted here must be the date when the Tender NSC/1 is signed as 'approved' by the Architect/Supervising Officer on behalf of the Employer

is made the ______ day of ______ 19 ______

between ______

of or whose registered office is situated at ______

______

(hereinafter called 'the Employer') and

______

of or whose registered office is situated at ______

______

(hereinafter called 'the Sub-Contractor')

Whereas

First — the Sub-Contractor has submitted a tender on Tender NSC/1 (hereinafter called 'the Tender') on the terms and conditions in that Tender to carry out Works (referred to above and hereinafter called 'the Sub-Contract Works') as part of the Main Contract Works referred to above to be or being carried out on the terms and conditions relating thereto referred to in Schedule 1 of the Tender (hereinafter called 'the Main Contract');

Second — the Employer has appointed

______

to be the Architect/Supervising Officer for the purposes of the Main Contract and this Agreement (hereinafter called 'the Architect/Supervising Officer' which expression as used in this Agreement shall include his successors validly appointed under the Main Contract or otherwise before the Main Contract is operative).

Third — the Architect/Supervising Officer on behalf of the Employer has approved the Tender and intends that after agreement between the Contractor and Sub-Contractor on the Particular Conditions in Schedule 2 thereof an instruction on Nomination NSC/3 shall be issued to the Contractor for the Main Contract (hereinafter called 'the Main Contractor') nominating the Sub-Contractor to carry out and complete the Sub-Contract Works on the terms and conditions of the Tender;

Fourth — nothing contained in this Agreement nor anything contained in the Tender is intended to render the Architect/Supervising Officer in any way liable to the Sub-Contractor in relation to matters in the said Agreement and Tender.

Page 1

Agreement NSC/2

Now it is hereby agreed

Completion of Tender: Sub-Contractor's obligations

1·1 The Sub-Contractor shall, after the Architect/Supervising Officer has issued his preliminary notice of nomination under clause 35·7·1 of the Main Contract Conditions, forthwith seek to settle with the Main Contractor the Particular Conditions in Schedule 2 of the Tender.

1·2 The Sub-Contractor shall, upon reaching agreement with the Main Contractor on the Particular Conditions in Schedule 2 of the Tender and after that Schedule is signed by or on behalf of the Sub-Contractor and the Main Contractor, immediately through the Main Contractor so inform the Architect/Supervising Officer.

Design, materials, performance specification

2·1 The Sub-Contractor warrants that he has exercised and will exercise all reasonable skill and care in

·1 the design of the Sub-Contract Works insofar as the Sub-Contract Works have been or will be designed by the Sub-Contractor; and

·2 the selection of materials and goods for the Sub-Contract Works insofar as such materials and goods have been or will be selected by the Sub-Contractor; and

·3 the satisfaction of any performance specification or requirement insofar as such performance specification or requirement is included or referred to in the description of the Sub-Contract Works included in or annexed to the Tender.

Nothing in clause 2·1 shall be construed so as to affect the obligations of the Sub-Contractor under Sub-Contract NSC/4 in regard to the supply under the Sub-Contract of workmanship, materials and goods.

2·2 ·1 If, after the date of this Agreement and before the issue by the Architect/Supervising Officer of the instruction on Nomination NSC/3 under clause 35·10·2 of the Main Contract Conditions, the Architect/Supervising Officer instructs in writing that the Sub-Contractor should proceed with
·1 the designing of, or
·2 the proper ordering or fabrication of any materials or goods for
the Sub-Contract Works the Sub-Contractor shall forthwith comply with the instruction and the Employer shall make payment for such compliance in accordance with clauses 2·2·2 to 2·2·4.

·2 No payment referred to in clauses 2·2·3 and 2·2·4 shall be made after the issue of Nomination NSC/3 under clause 35·10·2 of the Main Contract Conditions except in respect of any design work properly carried out and/or materials or goods properly ordered or fabricated in compliance with an instruction under clause 2·2·1 but which are not used for the Sub-Contract Works by reason of some written decision against such use given by the Architect/Supervising Officer before the issue of Nomination NSC/3.

·3 The Employer shall pay the Sub-Contractor the amount of any expense reasonably and properly incurred by the Sub-Contractor in carrying out work in the designing of the Sub-Contract Works and upon such payment the Employer may use that work for the purposes of the Sub-Contract Works but not further or otherwise.

·4 The Employer shall pay the Sub-Contractor for any materials or goods properly ordered by the Sub-Contractor for the Sub-Contract Works and upon such payment any materials and goods so paid for shall become the property of the Employer.

·5 If any payment has been made by the Employer under clauses 2·2·3 and 2·2·4 and the Sub-Contractor is subsequently nominated in Nomination NSC/3 issued under clause 35·10·2 of the Main Contract Conditions to execute the Sub-Contract Works the Sub-Contractor shall allow to the Employer and the Main Contractor full credit for such payment in the discharge of the amount due in respect of the Sub-Contract Works.

Delay in supply of information and in performance by Sub-Contractor

3·1 The Sub-Contractor will not be liable under clauses 3·2, 3·3 or 3·4 until the Architect/Supervising Officer has issued his instruction on Nomination NSC/3 under clause 35·10·2 of the Main Contract Conditions nor in respect of any revised period of time for delay in carrying out or completing the Sub-Contract Works which the Sub-Contractor has been granted under clause 11·2 of Sub-Contract NSC/4.

3.2 The Sub-Contractor shall so supply the Architect/Supervising Officer with information (including drawings) in accordance with the agreed programme details or at such time as the Architect/Supervising Officer may reasonably require so that the Architect/Supervising Officer will not be delayed in issuing necessary instructions or drawings under the Main Contract, for which delay the Main Contractor may have a valid claim to an extension of time for completion of the Main Contract Works by reason of the Relevant Event in clause 25·4·6 or a valid claim for direct loss and/or expense under clause 26·2·1 of the Main Contract Conditions.

3·3 The Sub-Contractor shall so perform his obligations under the Sub-Contract that the Architect/Supervising Officer will not by reason of any default by the Sub-Contractor be under a duty to issue an instruction to determine the employment of the Sub-Contractor under clause 35·24 of the Main Contract Conditions provided that any suspension by the Sub-Contractor of further execution of the Sub-Contract Works under clause 21·8 of Sub-Contract NSC/4 shall not be regarded as a 'default by the Sub-Contractor' as referred to in clause 3·3.

3·4 The Sub-Contractor shall so perform the Sub-Contract that the Contractor will not become entitled to an extension of time for completion of the Main Contract Works by reason of the Relevant Event in clause 25·4·7 of the Main Contract Conditions.

Architect's direction on value of Sub-Contract Work in Interim Certificates – information to Sub-Contractor

4 The Architect/Supervising Officer shall operate the provisions of *clause 35·13·1 of the Main Contract Conditions.

Employer's duty – final payment for Sub-Contract Works

5·1 The Architect/Supervising Officer shall operate the provisions in †clauses 35·17 to 35·19 of the Main Contract Conditions.

Discharge of final payment to Sub-Contractor – Sub-Contractor's obligations

5·2 After due discharge by the Contractor of a final payment under clause 35·17 of the Main Contract Conditions the Sub-Contractor shall rectify at his own cost (or if he fails so to rectify, shall be liable to the Employer for the costs referred to in clause 35·18 of the Main Contract Conditions) any omission, fault or defect in the Sub-Contract Works which the Sub-Contractor is bound to rectify under Sub-Contract NSC/4 after written notification thereof by the Architect/Supervising Officer at any time before the issue of the Final Certificate under clause 30·8 of the Main Contract Conditions.

5·3 After the issue of the Final Certificate under the Main Contract Conditions the Sub-Contractor shall in addition to such other responsibilities, if any, as he has under this Agreement, have the like responsibility to the Main Contractor and to the Employer for the Sub-Contract Works as the Main Contractor has to the Employer under the terms of the Main Contract relating to the obligations of the Contractor after the issue of the Final Certificate.

Architect's instructions – duty to make a further nomination – liability of Sub-Contractor

6 Where the Architect/Supervising Officer has been under a duty under clause 35·24 of the Main Contract Conditions except as a result of the operation of clause 35·24·6 to issue an instruction to the Main Contractor making a further nomination in respect of the Sub-Contract Works, the Sub-Contractor shall indemnify the Employer against any direct loss and/or expense resulting from the exercise by the Architect/Supervising Officer of that duty.

Architect's certificate of non-discharge by Contractor – payment to Sub-Contractor by Employer

7·1 The Architect/Supervising Officer and the Employer shall operate the provisions in regard to the payment of the Sub-Contractor in clause 35·13 of the Main Contract Conditions.

7·2 If, after paying any amount to the Sub-Contractor under clause 35·13·5·3 of the Main Contract Conditions, the Employer produces reasonable proof that there was in existence at the time of such payment a petition or resolution to which clause 35·13·5·4·4 of the Main Contract Conditions refers, the Sub-Contractor shall repay on demand such amount.

Conditions in contracts for purchase of goods and materials – Sub-Contractor's duty

8 Where ‡clause 2·3 of Sub-Contract NSC/4 applies, the Sub-Contractor shall forthwith supply to the Contractor details of any restriction, limitation or exclusion to which that clause refers as soon as such details are known to the Sub-Contractor.

Conflict between Tender and Agreement

9 If any conflict appears between the terms of the Tender and this Agreement, the terms of this Agreement shall prevail.

Arbitration

10·1 In case any dispute or difference shall arise between the Employer or the Architect/Supervising Officer on his behalf and the Sub-Contractor, either during the progress or after the completion or abandonment of the Sub-Contract Works, as to the construction of this Agreement, or as to any matter or thing of whatsoever nature arising out of this Agreement or in connection therewith, then such dispute or difference shall be and is hereby referred to the arbitration and final decision of a person to be agreed between the parties, or, failing agreement within 14 days after either party has given to the other a written request to concur in the appointment of an Arbitrator, a person to be appointed on the request of either party by the President or a Vice President for the time being of the Royal Institute of British Architects.

---

*Note: Clause 35·13·1 requires that the Architect/Supervising Officer, upon directing the Main Contractor as to the amount included in any Interim Certificates in respect of the value of the Nominated Sub-Contract Works issued under clause 30 of the Main Contract Conditions, shall forthwith inform the Sub-Contractor in writing of that amount.

†Note: Clause 35·17 deals with final payment by the Employer for the Sub-Contract Works prior to the issue of the Final Certificate under the Main Contract Conditions.

‡Note: Clause 2·3 deals with specified supplies and restrictions etc. in the contracts of sale for such supplies.

Agreement NSC/2

10·2 ·1 Provided that if the dispute or difference to be referred to arbitration under this Agreement raises issues which are substantially the same as or connected with issues raised in a related dispute between the Employer and the Contractor under the Main Contract or between the Sub-Contractor and the Contractor under Sub-Contract NSC/4 or NSC/4a or between the Employer and any other nominated Sub-Contractor under Agreement NSC/2 or NSC/2a or between the Employer and any Nominated Supplier whose contract of sale with the Main Contractor provides for the matters referred to in clause 36·4·8 of the Main Contract Conditions, and if the related dispute has already been referred for determination to an Arbitrator, the Employer and the Sub-Contractor hereby agree that the dispute or difference under this Agreement shall be referred to the Arbitrator appointed to determine the related dispute; and such Arbitrator shall have power to make such directions and all necessary awards in the same way as if the procedure of the High Court as to joining one or more defendants or joining co-defendants or third parties was available to the parties and to him.

10·2 ·2 Save that the Employer or the Sub-Contractor may require the dispute or difference under this Agreement to be referred to a different Arbitrator (to be appointed under this Agreement) if either of them reasonably considers that the Arbitrator appointed to determine the related dispute is not appropriately qualified to determine the dispute or difference under this Agreement.

10·2 ·3 Clauses 10·2·1 and 10·2·2 shall apply unless in the Appendix to the Main Contract Conditions the words "Articles 5·1·4 and 5·1·5 apply" have been deleted.

10·3 Such reference shall not be opened until after Practical Completion or alleged Practical Completion of the Main Contract Works or termination or alleged termination of the Contractor's employment under the Main Contract or abandonment of the Main Contract Works, unless with the written consent of the Employer or the Architect/Supervising Officer on his behalf and the Sub-Contractor.

10·4 The award of such Arbitrator shall be final and binding on the parties.

10·5* Whatever the nationality, residence or domicile of the Employer, the Contractor, any Sub-Contractor or supplier or the Arbitrator, and wherever the Works or any part thereof are situated, the law of England shall be the proper law of this Agreement and in particular (but not so as to derogate from the generality of the foregoing) the provisions of the Arbitration Acts 1950 (notwithstanding anything in S·34 thereof) to 1979 shall apply to any arbitration under Clause 10 wherever the same, or any part of it, shall be conducted.

*Where the parties do not wish the proper law of the contract to be the law of England and/or do not wish the provisions of the Arbitration Acts 1950 to 1979 to apply to any arbitration under the Contract held under the procedural law of Scotland (or other country) appropriate amendments to Clause 10·5 should be made.

Signed by or on behalf of the Sub-Contractor [g1] ______________________

in the presence of:

Signed by or on behalf of the Employer [g1] ______________________

in the presence of:

Signed, sealed and delivered by [g2]/The common seal of [g3]: ______________________

in the presence of [g2]/was hereunto affixed in the presence of [g3]:

Signed, sealed and delivered by [g2]/The common seal of [g3]: ______________________

in the presence of [g2]/was hereunto affixed in the presence of [g3]:

Footnotes

[g1] For use if Agreement is executed under hand.

[g2] For use if executed under seal by an individual or firm or unincorporated body.

[g3] For use if executed under seal by a company or other body corporate.

Architect's Instruction under Clause 35.11 nominating the sub-contractor if Tender NSC/1 is not used.
Both forms of Agreement are completed by the architect, signed by both employer and sub-contractor and dated immediately after signature. It is advisable to have the Agreement signed before nomination and return of NSC/1 by the contractor indicating he raises no objection to the nomination. Copies of the Agreement should be sent to both employer and sub-contractor, the original being held by the architect.
These forms relate only to the Standard Form of Building Contract.

Under the IFC (Clause 3.3.1) the architect may include in the specification/schedule of works/contract bills, the direction that works described shall be carried out by a named person to be employed as a sub-contractor by the contractor. The named person is required to tender on the Form of Tender and Agreement NAM/T which includes three sections:

- *Section I* Invitation to tender completed by the architect.
- *Section II* Tender by sub-contractor completed by the named person.
- *Section III* Articles of Agreement comprising the sub-contract agreement between named person and contractor, signed by both parties and witnessed.

The sub-contract conditions are those incorporated in Clause 1.2 of the Articles of Agreement set out in Section III of the Form of Tender and Agreement NAM/T and Sub-contract Conditions NAM/SC 1984 Edition.
Where a provisional sum is included, this procedure is also followed (Clause 3.3.1(c)). An instruction must be issued as to the expenditure of any provisional sums (Clause 3.8). To provide a direct contractual responsibility to the employer by the sub-contractor the RIBA/CASEC Form of Employer/Specialist Agreement ESA/1 can be used. This agreement is completed by the architect and signed by both employer and specialist (see pp. 72–77).
The placing of sub-contracts may be made in three ways:

- By means of a letter with copies to the employer, quantity surveyor and clerk of works (if employed).
- By the issue of an Architect's Instruction (see Fig 3.10) where the architect has elected under Clause 35.5.1.2 of the Standard Form of Building Contract not to use Tender Form NSC/1 and Agreement NSC/2 (see p. 67). If the contractor objects to the sub-contract he must object in writing within seven days from his receipt of the Instruction.

RIBA / CASEC Form of Employer / Specialist Agreement

for use between the Employer and a Specialist to be named under the JCT Intermediate Form of Building Contract (IFC 84)

Agreement ESA/1

1984 Edition

Guidance Note on RIBA/CASEC Form of Employer/Specialist Agreement for use between the Employer and a Specialist to be named as a Sub-Contractor under the JCT Intermediate Form of Building Contract.

1 The JCT Intermediate Form of Building Contract (IFC 84), referred to in this note as the 'Main Contract', provides for part of the work to be executed by a person named either in the tender documents for the Main Contract or in an instruction issued by the Architect for the expenditure of a provisional sum. The person named will then be employed by the Contractor as a sub-contractor subject to the provisions set out in Clauses 3·3·1 to 3·3·7 of the Main Contract.

2 Clause 3·3·7 of the Main Contract states that the Contractor shall not be responsible to the Employer under the Contract for a failure of such a sub-contractor to exercise reasonable care and skill in:

- the design of the sub-contract works insofar as the sub-contract works have been or will be designed by the named person;
- the selection of the kinds of materials and goods for the sub-contract works insofar as such materials and goods have been or will be selected by the named person; or
- the satisfaction of any performance specification or requirement relating to the sub-contract works.

The purpose of the RIBA/CASEC Form of Employer/Specialist Agreement, ESA/1, is to provide for such a sub-contractor to have a direct contractual responsibility to the Employer in respect of these matters.

3 The need for an Agreement must be considered according to the circumstances of each project and the degree of involvement of the Specialist. Although the Main Contractor is not a party to the Agreement he should be made aware of the contents of the Agreement in the Main Contract tender documents or before the issue of the Architect's instruction.

4 Whether the naming is in the tender documents or in an Architect's instruction, the Main Contract requires the use of the JCT Form of Tender and Agreement (NAM/T). This is in three sections. Section I, the Invitation to Tender, is to be completed by the Architect. Section II, the sub-contractor's Tender in response to the Invitation, is to be completed by the sub-contractor. The completion of Section III, Articles of Agreement, by the Contractor and the sub-contractor subsequently converts the Tender and Agreement into a sub-contract.

5 There are two ways in which ESA/1 can be used. When it is issued either paragraphs B and BB should be deleted (Procedure A, below) or, where these paragraphs are to apply, paragraphs A and AA should be deleted (Procedure B, below). In either case the requirements as stated in the Schedule are the same and paragraphs 7·1 and 7·2 are optional.

It is important that the actual periods required for the production of information referred to in the Schedule are noted, and the involvement of the Specialist may be required early in the pre-contract period if the Main Contractor is to be provided with the information required at the correct time. Where early ordering of materials or the commencement of fabrication may be required optional Clauses 7·1 and 7·2 should be included, although they will not come into operation unless specifically authorised by the Employer.

Procedure A:

When sufficient information is available for a final Tender to be requested from the Specialist

6 When paragraphs A and AA are to apply the procedure to be followed is:

6·1 Architect completes Section I of NAM/T.

6·2 Architect completes ESA/1 in part as follows:

·1 Insert names and addresses of Employer, Specialist and Architect on page 1.
·2 Delete paragraph B on page 1.
·3 In the definition of 'the Tender' in paragraph 6 on page 2, delete 'will be submitted by the Specialist to the Architect'.
·4 Delete paragraphs 7·1 and 7·2 on page 2 if not required.
·5 Insert period for acceptance of offer in paragraph AA on page 3.
·6 Delete paragraph BB on page 3.
·7 Delete provision for attestation by sealing on page 3 if not required.
Note:
If the Agreement is to be attested by the seals of both parties the paragraph stating that £10 shall be payable as consideration may be deleted.
·8 Complete Appendix on page 4.
Note:
When completing Part 1 of the Appendix it may be sufficient to refer to the information contained or referred to in Section I of NAM/T.

6·3 Architect sends ESA/1 and NAM/T, completed as above, to the Specialist who is being invited to tender.

6·4 Specialist completes Section II of NAM/T.

6·5 Specialist makes copy of NAM/T and marks the copy for the purposes of paragraph AA and annexes it to ESA/1.

6·6 Specialist completes ESA/1 as follows:

·1 Insert in paragraph AA the above reference to the copy of NAM/T which is annexed.
·2 Sign and date the offer in the space given on page 3.

6·7 Specialist returns ESA/1 and NAM/T, completed by him as above, to the Architect.

6·8 If the offer as made by the Specialist in ESA/1 is accepted, the Employer's signature and the date are to be inserted in the space given on page 3.
Note:
The £10 consideration (unless previously deleted, see note to 6·2·7 above) is then payable by the Employer to the Specialist or it may be agreed that the payment be made when the sub-contract has been entered into.

6·9 Where applicable, the attestation on page 3 of ESA/1 to be completed with the seals of the parties.

6·10 ESA/1 as so completed constitutes the Agreement between the Employer and the Specialist and is to be retained by or for the Employer in the same way as other original contract documents. A duplicate or certified copy of the Agreement should be passed to the Specialist.

Procedure B:

When sufficient information **is not** available for a final Tender to be requested from the Specialist.

7 Paragraphs A and AA cannot be used unless the Architect is already in a position to complete Section I of NAM/T.

When the design or specification will not be sufficiently advanced until the Specialist has commenced his design work, an approximate estimate should be obtained based on the restricted information then available. On the basis of this estimate the Specialist will proceed with the development of his design so as to enable a final Tender to be submitted. Unless the whole character of the work has changed there is no need for a revised Agreement to be entered into at this later stage. Procedure B is as follows:

7·1 Architect completes the form in part as follows:

·1 Insert names and addresses of Employer, Specialist and Architect on page 1.
·2 Delete paragraph A on page 1.
·3 In the definition of 'the Tender' in paragraph 6 on page 2 delete 'is identified on page 3 and annexed hereto'.
·4 Delete paragraphs 7·1 and 7·2 on page 2 if not required.
·5 Delete paragraph AA on page 3.
·6 Insert period for acceptance of offer in paragraph BB on page 3.
·7 Delete provision for attestation by sealing on page 3 if not required.
Note (a):
If the Agreement is to be attested by the seals of both parties the paragraph stating that £10 shall be payable as consideration may be deleted.
Note (b):
In addition to the consideration of £10 it is provided that if the sub-contract is not entered into by the Specialist the Employer will pay the abortive design costs incurred by the Specialist as described in paragraph BB of ESA/1.
·8 Complete Appendix on page 4.
Note:
The information referred to in Part 1 of the Appendix must be sufficient to enable the Specialist to prepare the approximate estimate which he is being invited to submit, including such as that indicated in Section I of NAM/T.

7·2 Architect sends ESA/1, completed as above, to the Specialist who is being invited to submit an approximate estimate.

7·3 Specialist prepares approximate estimate as requested.

7·4 Specialist marks the approximate estimate for the purposes of paragraph BB and annexes it to ESA/1.

7·5 Specialist completes ESA/1 as follows:

·1 Insert in paragraph BB the above reference to the approximate estimate.
·2 Sign and date the offer in the space given on page 3.

7·6 Specialist returns ESA/1, completed by him as above, with his approximate estimate annexed, to the Architect.

7·7 If the offer as made by the Specialist in ESA/1 is accepted, the Employer's signature and the date are to be inserted in the space given on page 3.
Note:
The £10 consideration (unless previously deleted, see note (a) to 7·1·7 above) is then payable by the Employer to the Specialist or it may be agreed that the payment be made when the sub-contract has been entered into.

7·8 Where applicable, the attestation on page 3 of ESA/1 to be completed with the seals of the parties.

7·9 ESA/1 as so completed constitutes the Agreement between the Employer and the Specialist and is to be retained by or for the Employer in the same way as other original contract documents. A duplicate or certified copy of the Agreement should be passed to the Specialist.

7·10 In due course the Architect completes Section I of NAM/T and sends it to the Specialist for the Specialist to submit his tender for the sub-contract works.

This form has been prepared by the Royal Institute of British Architects and the Committee of Associations of Specialist Engineering Contractors.

Published by RIBA Publications Limited
Finsbury Mission, Moreland Street
London EC1V 8VB

Printed by Duolith Limited
Welwyn Garden City, Herts

Agreement ESA/1

RIBA/CASEC Form of Employer/Specialist Agreement

for use between the Employer and a Specialist to be named as a sub-contractor under the JCT Intermediate Form of Building Contract 1984 Edition (IFC 84)

Between the Employer namely ______________________

____

____

and the Specialist namely ______________________

____

The Employer has appointed ______________________

____

as the Architect for the purposes hereof and has provided or caused to be provided to the Specialist the information referred to in the Appendix, part 1, including a brief description of the Sub-Contract Works hereinafter referred to.

INVITATION

The Specialist is hereby invited:

Delete either box A or box B

A	To submit herewith a tender for the Sub-Contract Works mentioned in paragraph 6 of the Schedule hereto ('the Tender') and to satisfy the requirements described or referred to in the Schedule hereto ('the Requirements').
B	To submit herewith an approximate estimate in respect of the Sub-Contract Works mentioned in paragraph 6 of the Schedule hereto ('the Approximate Estimate') and to satisfy the requirements described or referred to in the Schedule hereto ('the Requirements'), which include the submission of a tender ('the Tender').

OFFER AND AGREEMENT

Schedule referred to above

1 The Requirements of the Employer described or referred to herein relate to the Sub-Contract Works. Save to the extent otherwise agreed, the time requirements are to be satisfied by the Specialist subject to the Employer providing or causing to be provided to the Specialist the further information (if any) referred to in the Appendix, part 2, at the time or times therein prescribed.

2 The information required to be provided by the Specialist to the Architect shall also be provided so as to enable the Architect to co-ordinate and integrate the design of the Sub-Contract Works into the design for the Main Contract Works as a whole.

3 The Specialist is required to provide to the Architect:

3·1 according to such time requirements as are stated in the Appendix, part 3, the information required for the Tender to be used, as the case may be

either
– for the purposes of (a) obtaining for the Employer tenders for the Main Contract Works and (b) inclusion in the contract documents of the main contract;

or
– for the purposes of enabling the Architect to issue an instruction to the main contractor as to the expenditure of a provisional sum requiring the Sub-Contract Works to be executed by the Specialist as a sub-contractor employed by the main contractor;

Footnote

This Agreement does not form part of the Agreement between the Employer and the Main Contractor or of that between the Main Contractor and the Specialist. While it may be sent to the Specialist with the Form of Tender and Agreement NAM/T for him to complete and return, it should not be included in the main contract documents or in an instruction of the Architect for the expenditure of a Provisional Sum; it should be retained for signing/sealing by the Employer.

Page 1

Agreement ESA/1

OFFER AND AGREEMENT
continued

3·2 according to the time requirements stated or referred to in the Tender and/or the Sub-Contract, such further information relating to the Sub-Contract Works as is reasonably necessary to enable the Architect to provide the main contractor with such information as is reasonably necessary to enable the main contractor:

— to carry out and complete the Main Contract Works in accordance with the conditions of the main contract, including the Sub-Contract Works to be executed by the Specialist as the Sub-Contractor; and

— to provide the Specialist in accordance with the conditions of the Sub-Contract with such information as is reasonably necessary to enable the Specialist to carry out and complete the Sub-Contract Works in accordance with the Sub-Contract.

4 The Employer shall be entitled to use for the purposes of carrying out and completing the Main Contract Works and maintaining or altering the Main Contract Works any drawings or information provided by the Specialist in accordance with this Agreement.

5 The Requirements include the exercise by the Specialist of reasonable care and skill in:

— the design of the Sub-Contract Works insofar as the Sub-Contract Works have been or will be designed by the Specialist; and

— the selection of materials and goods for the Sub-Contract Works insofar as such materials and goods have been or will be selected by the Specialist; and

— the satisfaction of any performance specification or requirement included or referred to in the description of the Sub-Contract Works included in or annexed to the Tender.

Definitions

6 Whether or not the Specialist's tender for the Sub-Contract Works is accepted and a Sub-Contract entered into, in this Offer and Agreement:

— 'the Architect' *means* the Architect named on page 1 as appointed by the Employer or such other person who is named in Section 1 of the Tender as 'the Architect' or as 'the Supervising Officer' for the main contract or such other person as the Employer appoints in place of the person so named;

— 'the Form of Tender NAM/T' *means* the form issued by the Joint Contracts Tribunal for use with the JCT Intermediate Form of Building Contract;

**Delete as appropriate*

and 'the Tender' *means* such a form as completed for the purposes of the Specialist's tender (a copy of which * is identified on page 3 and annexed hereto/* will be submitted by the Specialist to the Architect) containing (a) information given to the Specialist about the main contract for which the Sub-Contract Works are required, and (b) information about the proposed conditions of the sub-contract to be entered into between the Specialist and the main contractor for the execution of the Sub-Contract Works;

— 'information' *includes*, wherever appropriate, drawings and the information to be submitted with the Approximate Estimate and/or the Tender;

— 'the Main Contract Works' *means* the works of which the Sub-Contract Works form a part and which are described in Section 1 of the Tender;

— 'the Sub-Contract Works' *means* the work to be executed by the Specialist as a sub-contractor named under the provisions of a building contract between the Employer and a main contractor, after receipt of the Specialist's tender using the Form of Tender NAM/T;

and 'the Sub-Contract' *means* the Sub-Contract entered into by the Specialist as such named person with the main contractor.

Materials and Goods

Delete paragraphs 7·1 and 7·2 if not applicable

7·1 After the date of this Agreement but before the Specialist enters into a sub-contract for the execution of the Sub-Contract Works the Employer may require the Specialist to proceed with the purchase of materials or goods or the fabrication of components for the Sub-Contract Works.

7·2 If the Sub-Contract is not entered into then, subject to any agreement to the contrary, the Employer shall pay the Specialist for any such materials and goods which have been purchased or components properly fabricated whereupon they shall become the property of the Employer.

Page 2

Agreement ESA/1

OFFER AND AGREEMENT
continued

Delete either box AA or box BB

AA A copy of the Tender referred to in A on page 1 is **annexed hereto marked**

____

We offer to satisfy the Requirements described or referred to in the Schedule above in accordance with this Agreement.

This offer is withdrawn if not accepted within __________ weeks of the date of our signature below.

BB The Approximate Estimate referred to in B on page 1 is **annexed hereto marked**

____

We offer to satisfy the Requirements described or referred to in the Schedule above in accordance with this Agreement:

— **Provided** that if the Sub-Contract is not entered into by the Specialist, the Employer shall pay the Specialist the amount of any expenses reasonably and properly incurred by the Specialist in carrying out work in the designing of the Sub-Contract Works in anticipation of the Sub-Contract and in accordance with the Requirements.

This offer is withdrawn if not accepted within __________ weeks of the date of our signature below.

In consideration whereof the sum of £10 (plus VAT as appropriate) shall be payable to the Specialist.

Signed ______________________________ Dated __________

on behalf of the Specialist

This offer is accepted

Signed ______________________________ Dated __________

(on behalf of) the Employer

In witness of this Agreement the parties have hereunto set their seals

this __________ day of __________ 19 __________

Signed, sealed and delivered by:

in the presence of:

Or The common seal of:

was hereunto affixed in the presence of:

Signed, sealed and delivered by:

in the presence of:

Or The common seal of:

was hereunto affixed in the presence of:

continued

Agreement ESA/1

Appendix

Referred to in the Form of Agreement

Part 1 Information provided herewith by the Employer (referred to on page 1)

Part 2 Further information which the Employer is to provide or cause to be provided, referred to in paragraph 1 of the Schedule

Information	Date

Part 3 Time requirements referred to in paragraph 3·1 of the Schedule

For Approximate Estimate (if applicable)

For Tender

This form has been prepared by the Royal Institute of British Architects and the Committee of Associations of Specialist Engineering Contractors.

Page 4

- By the architect sending Tender Form NSC/1 and any related documents duly completed by the proposed sub-contractor to the contractor for completion and return to the architect. On receipt of the completed Form NSC/1 the architect nominates the sub-contractor on the Standard Form of Nomination of Sub-contractor, where Tender NSC/1 has been used, sent direct to the contractor with all necessary supporting documentation relating to the sub-contract with copies to the sub-contractor concerned and relevant members of the design team. This method is used under Clause 35.4.2 of the Standard Form of Building Contract. If the Contractor objects to the nomination he must do so when he returns the completed Tender Form NSC/1 to the architect.

Where Tender form NSC/1 has been used the contractor enters into a sub-contract with the nominated sub-contractor using Sub-contract NSC/4. Where the tender form has not been used the contractor uses Sub-contract NSC/4a. The Tender Form NSC/1 incorporates a 2½ per cent cost discount allowable to the contractor and this discount is incorporated into Sub-contract NSC/4. The sub-contractor must have his attention drawn to this discount if NSC/1 is not used, the discount being deducted by the contractor on payments made to the sub-contractor within fourteen days after receipt by the contractor of the relevant architect's Certificate.

The nomination of suppliers for goods and materials may be made in two ways:

- By letter as above for nominated sub-contractors.
- By Architect's Instruction in the manner indicated in Fig 3.10.

Under Clause 36 of the Standard Form of Building Contract the conditions of nomination of suppliers in respect of goods and materials are defined. Such suppliers are required to enter into a contract of sale with the contractor against provision made in the contract, bills of quantities or specification by the inclusion of prime cost sums. The supplier must include for and allow in his tender a discount of 5 per cent to the contractor for payment within thirty days of the end of the month during which delivery is made.

With the use of the Standard Form of Building Contract tenders for nominated supplier are invited on form Tender TNS/1 (see pp. 86–89). The architect completes page 1 as required and Schedule 1 on pages 2 and 3 and attaches it to a schedule of materials required before sending the form to the relevant suppliers.

Fig 3.10 Architect's Instruction, from architect to contractor: instruction regarding prime cost sums

Prime cost sums

Nominated sub-contractor

[33] Omit PC sum of [£...] included in the bill of quantities specification for supply and fixing of fireproof steel doors.

Add Enter into a sub-contract with Messrs [*insert names*] for the supply, delivery to site and fixing complete of 2 no. fireproof steel doors on the basis of their tender dated [*insert date*] amounting to [£...], which sum includes 2½ per cent contractor's discount.

To enable you to do so we enclose the following:

Two copies of the relevant drawings and specification for sub-contract works (or schedules of work as applicable).

Two copies of the relevant tender.

The remainder of the form including the offer is completed by the tenderer. In due course the successful supplier will be nominated by the issue of an Architect's Instruction issued to the contractor accompanied by a copy of the Tender and associated documentation (see Fig 3.11). A copy of the nomination instruction etc. is also sent to the supplier.
The tender will include a cash discount of 5 per cent allowable to the contractor on the same terms of payment as for sub-contractors. This form is inapplicable for use with the Agreement for Minor Building Works, which does not envisage the use of sub-contractors under this contract.
A copy of the Form of Warranty is completed and attached to each invitation to tender to prospective nominated suppliers (see pp. 91–92). The form is signed and dated by the tenderer and returned with his tender.

Agreement NSC/2a

JCT

JCT Standard Form of Employer/Nominated Sub-Contractor Agreement

Agreement between a Sub-Contractor to be nominated for Sub-Contract Work in accordance with clauses 35·11 and 35·12 of the Standard Form of Building Contract for a main contract and the Employer referred to in the main contract.

Main Contract Works:

Location:

Sub-Contract Works:

This Agreement

The date to be inserted here must be a date not later than the date of the nomination instruction of the Architect/Supervising Officer under clause 35·11 of the Main Contract Conditions

is made the ______________ day of ______________ 19 ______

between ______________

of or whose registered office is situated at ______________

____

(hereinafter called 'the Employer') and

____

of or whose registered office is situated at ______________

____

(hereinafter called 'the Sub-Contractor')

Whereas

Identify edition and revision, if any, of the Standard Form

First — the Employer has entered into/intends to enter into* a contract (hereinafter called 'the Main Contract') for certain works (hereinafter called 'the Main Contract Works') and for which the Main Contract Conditions are:

____

(hereinafter called 'the Main Contract Conditions');

Second — the Employer has appointed

____

to be the Architect/Supervising Officer for the purposes of the Main Contract and this Agreement (hereinafter called 'the Architect/Supervising Officer' which expression as used in this Agreement shall include his successor validly appointed under the Main Contract or otherwise before the Main Contract is operative).

Third — the Sub-Contractor has tendered for certain works (hereinafter called 'the Sub-Contract Works') to be carried out by a Nominated Sub-Contractor as part of the Main Contract Works and the tender of the Sub-Contractor has been made on the basis that the Sub-Contractor and the Employer will enter into Agreement NSC/2a;

Fourth — the Architect/Supervising Officer on behalf of the Employer has approved the tender of the Sub-Contractor and the Main Contract (either in the Contract Bills forming part of the Main Contract or in an instruction of the Architect/Supervising Officer under the Main Contract Conditions) provides that clauses 35·11 and 35·12 of the Main Contract Conditions shall apply to the nomination of the Sub-Contractor;

Fifth — the Architect/Supervising Officer has issued/intends to issue* an instruction under clause 35·11 of the Main Contract Conditions nominating the Sub-Contractor to carry out and complete the Sub-Contract Works on the terms and conditions of Sub-Contract NSC/4a as referred to in clause 35·12 of the Main Contract Conditions;

Sixth — nothing contained in this Agreement is intended to render the Architect/Supervising Officer in any way liable to the Sub-Contractor in relation to matters in the said Agreement.

*Delete as applicable

Agreement NSC/2a

Now it is hereby agreed

Design, materials, performance specification

1·1 The Sub-Contractor warrants that he has exercised and will exercise all reasonable skill and care in

·1 the design of the Sub-Contract Works insofar as the Sub-Contract Works have been or will be designed by the Sub-Contractor; and

·2 the selection of materials and goods for the Sub-Contract Works insofar as such materials and goods have been or will be selected by the Sub-Contractor; and

·3 the satisfaction of any performance specification or requirement insofar as such performance specification or requirement is included or referred to in the Sub-Contract Documents as defined in Sub-Contract NSC/4a.

Nothing in clause 1·1 shall be construed so as to affect the obligations of the Sub-Contractor under Sub-Contract NSC/4a in regard to the supply under the Sub-Contract of workmanship, materials and goods.

1·2 ·1 If, after the date of this Agreement and before the issue by the Architect/Supervising Officer of the instruction under clause 35·11 of the Main Contract Conditions, the Architect/Supervising Officer instructs in writing that the Sub-Contractor should proceed with
·1 the designing of, or
·2 the proper ordering or fabrication of any materials or goods for
the Sub-Contract Works the Sub-Contractor shall forthwith comply with the instruction and the Employer shall make payment for such compliance in accordance with clauses 1·2·2 to 1·2·4.

·2 No payment referred to in clauses 1·2·3 and 1·2·4 shall be made after the issue of the instruction nominating the Sub-Contractor under clause 35·11 of the Main Contract Conditions except in respect of any design work properly carried out and/or materials or goods properly ordered or fabricated in compliance with an instruction under clause 1·2·1 but which are not used for the Sub-Contract Works by reason of some written decision against such use given by the Architect/Supervising Officer before the issue of the instruction nominating the Sub-Contractor under clause 35·11 of the Main Contract Conditions.

·3 The Employer shall pay the Sub-Contractor the amount of any expense reasonably and properly incurred by the Sub-Contractor in carrying out work in the designing of the Sub-Contract Works and upon such payment the Employer may use that work for the purposes of the Sub-Contract Works but not further or otherwise.

·4 The Employer shall pay the Sub-Contractor for any materials or goods properly ordered by the Sub-Contractor for the Sub-Contract Works and upon such payment any materials and goods so paid for shall become the property of the Employer.

·5 If any payment has been made by the Employer under clauses 1·2·3 and 1·2·4 and the Sub-Contractor is subsequently nominated by an instruction nominating the Sub-Contractor issued under clause 35·11 of the Main Contract Conditions to execute the Sub-Contract Works the Sub-Contractor shall allow to the Employer and the Main Contractor full credit for such payment in the discharge of the amount due in respect of the Sub-Contract Works.

Delay in supply of information and in performance by Sub-Contractor

2·1 The Sub-Contractor will not be liable under clauses 2·2, 2·3 or 2·4 until the Architect/Supervising Officer has issued his instruction nominating the Sub-Contractor issued under clause 35·11 of the Main Contract Conditions nor in respect of any revised period of time for delay in carrying out or completing the Sub-Contract Works which the Sub-Contractor has been granted under clause 11·2 of Sub-Contract NSC/4a.

2.2 The Sub-Contractor shall so supply the Architect/Supervising Officer with information (including drawings) in accordance with the agreed programme details or at such time as the Architect/Supervising Officer may reasonably require so that the Architect/Supervising Officer will not be delayed in issuing necessary instructions or drawings under the Main Contract, for which delay the Main Contractor may have a valid claim to an extension of time for completion of the Main Contract Works by reason of the Relevant Event in clause 25·4·6 or a valid claim for direct loss and/or expense under clause 26·2·1 of the Main Contract Conditions.

2·3 The Sub-Contractor shall so perform his obligations under the Sub-Contract that the Architect/Supervising Officer will not by reason of any default by the Sub-Contractor be under a duty to issue an instruction to determine the employment of the Sub-Contractor under clause 35·24 of the Main Contract Conditions provided that any suspension by the Sub-Contractor of further execution of the Sub-Contract Works under clause 21·8 of Sub-Contract NSC/4a shall not be regarded as a 'default by the Sub-Contractor' as referred to in clause 2·3.

2·4 The Sub-Contractor shall so perform the Sub-Contract that the Contractor will not become entitled to an extension of time for completion of the Main Contract Works by reason of the Relevant Event in clause 25·4·7 of the Main Contract Conditions.

Page 2

Agreement NSC/2a

Architect's direction on value of Sub-Contract Work in Interim Certificates – information to Sub-Contractor

3 The Architect/Supervising Officer shall operate the provisions of *clause 35·13·1 of the Main Contract Conditions.

Employer's duty – final payment for Sub-Contract Works

4·1 The Architect/Supervising Officer shall operate the provisions in †clauses 35·17 to 35·19 of the Main Contract Conditions.

Discharge of final payment to Sub-Contractor – Sub-Contractor's obligations

4·2 After due discharge by the Contractor of a final payment under clause 35·17 of the Main Contract Conditions the Sub-Contractor shall rectify at his own cost (or if he fails so to rectify, shall be liable to the Employer for the costs referred to in clause 35·18 of the Main Contract Conditions) any omission, fault or defect in the Sub-Contract Works which the Sub-Contractor is bound to rectify under Sub-Contract NSC/4a after written notification thereof by the Architect/Supervising Officer at any time before the issue of the Final Certificate under clause 30·8 of the Main Contract Conditions.

4·3 After the issue of the Final Certificate under the Main Contract Conditions the Sub-Contractor shall in addition to such other responsibilities, if any, as he has under this Agreement, have the like responsibility to the Main Contractor and to the Employer for the Sub-Contract Works as the Main Contractor has to the Employer under the terms of the Main Contract relating to the obligations of the Contractor after the issue of the Final Certificate.

Architect's instructions – duty to make a further nomination – liability of Sub-Contractor

5 Where the Architect/Supervising Officer has been under a duty under clause 35·24 of the Main Contract Conditions except as a result of the operation of clause 35·24·6 to issue an instruction to the Main Contractor making a further nomination in respect of the Sub-Contract Works, the Sub-Contractor shall indemnify the Employer against any direct loss and/or expense resulting from the exercise by the Architect/Supervising Officer of that duty.

Architect's certificate of non-discharge by Contractor – payment to Sub-Contractor by Employer

6·1 The Architect/Supervising Officer and the Employer shall operate the provisions in regard to the payment of the Sub-Contractor in clause 35·13 of the Main Contract Conditions.

6·2 If, after paying any amount to the Sub-Contractor under clause 35·13·5·3 of the Main Contract Conditions, the Employer produces reasonable proof that there was in existence at the time of such payment a petition or resolution to which clause 35·13·5·4·4 of the Main Contract Conditions refers, the Sub-Contractor shall repay on demand such amount.

Conditions in contracts for purchase of goods and materials – Sub-Contractor's duty

7 Where ‡clause 2·3 of Sub-Contract NSC/4a applies, the Sub-Contractor shall forthwith supply to the Contractor details of any restriction, limitation or exclusion to which that clause refers as soon as such details are known to the Sub-Contractor.

Arbitration

8·1 In case any dispute or difference shall arise between the Employer or the Architect/Supervising Officer on his behalf and the Sub-Contractor, either during the progress or after the completion or abandonment of the Sub-Contract Works, as to the construction of this Agreement, or as to any matter or thing of whatsoever nature arising out of this Agreement or in connection therewith, then such dispute or difference shall be and is hereby referred to the arbitration and final decision of a person to be agreed between the parties, or, failing agreement within 14 days after either party has given to the other a written request to concur in the appointment of an Arbitrator, a person to be appointed on the request of either party by the President or a Vice President for the time being of the Royal Institute of British Architects.

8·2 ·1 Provided that if the dispute or difference to be referred to arbitration under this Agreement raises issues which are substantially the same as or connected with issues raised in a related dispute between the Employer and the Contractor under the Main Contract or between the Sub-Contractor and the Contractor under Sub-Contract NSC/4 or NSC/4a or between the Employer and any other nominated Sub-Contractor under Agreement NSC/2 or NSC/2a or between the Employer and any Nominated Supplier whose contract of sale with the Main Contractor provides for the matters referred to in clause 36·4·8 of the Main Contract Conditions, and if the related dispute has already been referred for determination to an Arbitrator, the Employer and the Sub-Contractor hereby agree that the dispute or difference under this Agreement shall be referred to the Arbitrator appointed to determine the related dispute; and such Arbitrator shall have power to make such directions and all necessary awards in the same way as if the procedure of the High Court as to joining one or more defendants or joining co-defendants or third parties was available to the parties and to him.

8·2 ·2 Save that the Employer or the Sub-Contractor may require the dispute or difference under this Agreement to be referred to a different Arbitrator (to be appointed under this Agreement) if either of them reasonably considers that the Arbitrator appointed to determine the related dispute is not appropriately qualified to determine the dispute or difference under this Agreement

*Note: Clause 35·13·1 requires that the Architect/Supervising Officer, upon directing the Main Contractor as to the amount included in any Interim Certificates in respect of the value of the Nominated Sub-Contract Works issued under clause 30 of the Main Contract Conditions, shall forthwith inform the Sub-Contractor in writing of that amount.

†Note: Clause 35·17 deals with final payment by the Employer for the Sub-Contract Works prior to the issue of the Final Certificate under the Main Contract Conditions.

‡Note: Clause 2·3 deals with specified supplies and restrictions etc. in the contracts of sale for such supplies.

Agreement NSC/2a

8·2 ·3 Clauses 8·2·1 and 8·2·2 shall apply unless in the Appendix to the Main Contract Conditions the words 'Articles 5·1·4 and 5·1·5 apply' have been deleted.

8·3 Such reference shall not be opened until after Practical Completion or alleged Practical Completion of the Main Contract Works or termination or alleged termination of the Contractor's employment under the Main Contract or abandonment of the Main Contract Works, unless with the written consent of the Employer or the Architect/Supervising Officer on his behalf and the Sub-Contractor.

8·4 The award of such Arbitrator shall be final and binding on the parties.

8·5* Whatever the nationality, residence or domicile of the Employer, the Contractor, any Sub-Contractor or supplier or the Arbitrator, and wherever the Works or any part thereof are situated, the law of England shall be the proper law of this Agreement and in particular (but not so as to derogate from the generality of the foregoing) the provisions of the Arbitration Acts 1950 (notwithstanding anything in S·34 thereof) to 1979 shall apply to any arbitration under Clause 8 wherever the same, or any part of it, shall be conducted.

*Where the parties do not wish the proper law of the contract to be the law of England and/or do not wish the provisions of the Arbitration Acts 1950 to 1979 to apply to any arbitration under the Contract held under the procedural law of Scotland (or other country) appropriate amendments to Clause 8·5 should be made.

Signed by or on behalf of the Sub-Contractor [g1] ______________________

in the presence of:

Signed by or on behalf of the Employer [g1] ______________________

in the presence of:

Signed, sealed and delivered by [g2]/The common seal of [g3]: ______________________

in the presence of [g2]/was hereunto affixed in the presence of [g3]:

Signed, sealed and delivered by [g2]/The common seal of [g3]: ______________________

in the presence of [g2]/was hereunto affixed in the presence of [g3]:

Footnotes

[g1] For use if Agreement is executed under hand.

[g2] For use if executed under seal by an individual or firm or unincorporated body.

[g3] For use if executed under seal by a company or other body corporate.

Page 4

Nomination NSC/3

JCT

JCT Standard Form of Nomination of Sub-Contractor where Tender NSC/1 has been used

To: ____________________

____

(Main Contractor)

Main Contract Works: ____________________

____________________ Job reference: ____________________

Sub-Contract Works: ____________________

____

Page Number – Bills of Quantities or Specification: ____________________

Sub-Contractor hereby nominated: ____________________

of: ____________________

____

____________________ Tel. No: ____________________

Further to my/our Preliminary Notice of Nomination (Main Contract Conditions clause 35·7·2)

dated 19

and Tender NSC/1 (and annexed documents) duly completed by you and the Sub-Contractor named above and by myself/ourselves on behalf of the Employer, the Sub-Contractor named above is hereby **NOMINATED** under the Main Contract Conditions clause 35·10·2 for the Sub-Contract Works identified above.

Signed ____________________ Architect/Supervising Officer

Address ____________________

____

____

Date 19

Circulation

- ☐ Main Contractor
- ☐ Quantity Surveyor
- ☐ Sub-Contractor hereby nominated
- ☐ Clerk of Works
- ☐ Consulting Engineer
- ☐ Architect/Supervising Officer's file

Fig 3.11 Architect's Instruction, from architect to contractor: instruction regarding prime cost or provisional sums

Nominated supplier

[34] Omit PC sum of [£....] included in the bill of quantities specification for the supply and delivery to site of sanitary fittings.

Add Place an order with Messrs [*insert names*] for the supply and delivery to site of sanitary fittings on the basis of their tender dated [*insert date*] amounting to [£....] which sum includes 5 per cent contractor's discount.

To enable you to do so we enclose the following:

Two copies of the relevant schedules of materials on which the tender was based.

Two copies of the relevant tender.

Provisional sum

[35] Omit the provisional sum of [£....] for contingencies.

The warranty being between the employer and the supplier, the form is addressed to the employer and describes the main contract works for which the tender forms part. It is necessary to include the amount for payment in respect of liquidated damages and a short description of the sub-contract goods. The name of the tenderer is filled in after 'We' and Clause (4) is deleted or the percentage filled in as applicable.

The warranty form is kept attached to the tender by the architect on behalf of the employer.

Tender TNS/1

JCT Standard Form of Tender by Nominated Supplier

For use in connection with the Standard Form of Building Contract (SFBC) issued by the Joint Contracts Tribunal, 1980 edition, current revision

Job Title:
(name and brief location of Works)

Employer:[a]

Main Contractor:[a]
(if known)

Tender for:[a]
(abbreviated description)

Name of Tenderer:

To be returned to:[a]

Lump sum price:[b] £ ______________________________

______________________________ *(words)*

and/or Schedule of rates (attached)

1 We confirm that we will be under a contract with the Main Contractor:

·1 to supply the materials or goods described or referred to in **Schedule 1** for the price and/or at the rate set out above; and

·2 in accordance with the other terms set out in that Schedule, as a Nominated Supplier in accordance with the terms of SFBC clause 36·3 to ·5 (as set out in **Schedule 2**) and our conditions of sale in so far as they do not conflict with the terms of SFBC clause 36·3 to ·5 [c]

provided:

·3 the Architect/Supervising Officer has issued the relevant nomination instruction (a copy of which has been sent to us by the Architect/Supervising Officer); and

·4 agreement on delivery between us and the Main Contractor has been reached as recorded in **Schedule 1** Part 6 (see SFBC clause 36·4·3); and

·5 we have thereafter received an order from the Main Contractor accepting this tender.

2 We agree that this Tender shall be open for acceptance by an order from the Main Contractor within [d] of the date of this Tender. Provided that where the Main Contractor has not been named above we reserve the right to withdraw this Tender within 14 days of having been notified, by or on behalf of the Employer named above, of the name of the Main Contractor.

3[e] Subject to our right to withdraw this Tender as set out in paragraph 2 we hereby declare that we accept the Warranty Agreement in the terms set out in **Schedule 3** hereto on condition that no provision in that Warranty Agreement shall take effect unless and until

a copy to us of the instruction nominating us,
the order of the Main Contractor accepting this Tender, and
a copy of the Warranty Agreement signed by the Employer

have been received by us.

For and on behalf of

Address

Signature   Date

[a] To be completed by or on behalf of the Architect/Supervising Officer.

[b] To be completed by the supplier; see also Schedule 1, item 7.

[c] By SFBC clause 36·4·9 none of the provisions in the contract of sale can override, modify or affect in any way the provisions incorporated from SFBC clause 36·4 in that contract of sale. Nominated Suppliers should therefore take steps to ensure that their sale conditions do not incorporate any provisions which purport to override, modify or affect in any way the provisions incorporated from SFBC clause 36·4.

[d] May be completed by or on behalf of the Architect/Supervising Officer; if not so completed, to be completed by the supplier.

[e] To be struck out by or on behalf of the Architect/Supervising Officer if no Warranty Agreement is required.

Page 1

Tender TNS/1

Schedule 1

1. Description, quantity and quality of materials or goods:

1A

1B

1C

Note: 1A to be completed by or on behalf of the Architect/Supervising Officer setting out his requirements. If the supplier is unable to comply with 1A he is to state in 1B what modifications he proposes, and the Architect/Supervising Officer is to state in 1C if such modifications are acceptable.

2. Access to Works:

2A

2B

Note: 2A to be completed by or on behalf of the Architect/Supervising Officer. The supplier in 2B **either** *confirms that the access in 2A is acceptable* **or** *states what modifications etc. to the access he requires* **or** *if 2A has not been completed, completes 2B.*

3. Provisions, if any, for returnable packings:

4. Completion Date of Main Contract (or anticipated Completion Date if Main Contract not let):

Note: To be completed by or on behalf of the Architect/Supervising Officer.

5. Defects Liability Period of the Main Contract months.

Note: To be completed by or on behalf of the Architect/Supervising Officer.

Page 2

Tender TNS/1

6A Anticipated commencement and completion dates for the nominated supply after any necessary approval of drawings (subject to SFBC clause 36·4·3, which provides that delivery shall be commenced, carried out and completed in accordance with any delivery programme agreed between the Contractor and supplier or in the absence of such programme in accordance with the reasonable directions of the Contractor):

6B ·1 Supplier's proposed delivery programme to comply with 6A:

·2 If 6B·1 not completed, delivery programme shall be to the reasonable directions of the Contractor.

6C Delivery programme as agreed between the supplier and the Contractor, if different from 6B:

Note: 6A to be completed by or on behalf of the Architect/Supervising Officer. The supplier to complete 6B·1 to take account of 6A; but the completion of 6B·1 is subject to the terms of 6C which may need to be used when the Contractor and supplier are settling item 6.

7. Provisions, if any, for fluctuations in price or rates:

7A

7B

7C

Note: 7A to be completed by or on behalf of the Architect/Supervising Officer. If the supplier is unable to comply with 7A he is to state in 7B what modifications to the provisions in 7A he requires, and the Architect/Supervising Officer is to state in 7C if such modifications are acceptable.

8. SFBC clause 25 (extensions of time) applies to the Main Contract without modification except as stated below:

The liquidated and ascertained damages (SFBC clause 2·4·2 and Appendix entry) under the Main Contract are at the rate of £ per .

Note: To be completed by or on behalf of the Architect/Supervising Officer.

9. Contract of sale with Contractor to be under hand/under seal.

Note: Alternative not to be used to be deleted by the supplier subject to agreement on the method of execution of the sale contract with the Main Contractor.

Page 3

Tender TNS/1

Schedule 2

JCT Standard Form of Building Contract

Clause 36·3 to ·8 provides as follows:

Ascertainment of costs to be set against prime cost sum

36·3 ·1 For the purposes of clause 30·6·2·8 the amounts "properly chargeable to the Employer in accordance with the nomination instruction of the Architect/Supervising Officer" shall include the total amount paid or payable in respect of the materials or goods less any discount other than the discount referred to in clause 36·4·4·, properly so chargeable to the Employer and shall include where applicable:

·1·1 any tax (other than any value added tax which is treated, or is capable of being treated, as input tax (as referred to in the Finance Act 1972) by the Contractor) or duty not otherwise recoverable under this Contract by whomsoever payable which is payable under or by virtue of any Act of Parliament on the import, purchase, sale, appropriation, processing, alteration, adapting for sale or use of the materials or goods to be supplied; and

·1·2 the net cost of appropriate packing carriage and delivery after allowing for any credit for return of any packing to the supplier; and

·1·3 the amount of any price adjustment properly paid or payable to, or allowed or allowable by the supplier less any discount other than a cash discount for payment in full within 30 days of the end of the month during which delivery is made.

·2 Where in the opinion of the Architect/Supervising Officer the Contractor properly incurs expense, which would not be reimbursed under clause 36·3·1 or otherwise under this Contract, in obtaining the materials or goods from the Nominated Supplier such expense shall be added to the Contract Sum.

Sale contract provisions – Architect's/Supervising Officer's right to nominate supplier

36·4 Save where the Architect/Supervising Officer and the Contractor shall otherwise agree, the Architect/Supervising Officer shall only nominate as a supplier a person who will enter into a contract of sale with the Contractor which provides, inter alia:

·1 that the materials or goods to be supplied shall be of the quality and standard specified provided that where and to the extent that approval of the quality of materials or of the standards of workmanship is a matter for the opinion of the Architect/Supervising Officer, such quality and standards shall be to the reasonable satisfaction of the Architect/Supervising Officer;

·2 that the Nominated Supplier shall make good by replacement or otherwise any defects in the materials or goods supplied which appear up to and including the last day of the Defects Liability Period under this Contract and shall bear any expenses reasonably incurred by the Contractor as a direct consequence of such defects provided that:

·2·1 where the materials or goods have been used or fixed such defects are not such that reasonable examination by the Contractor ought to have revealed them before using or fixing;

·2·2 such defects are due solely to defective workmanship or material in the materials or goods supplied and shall not have been caused by improper storage by the Contractor or by misuse or by any act or neglect of either the Contractor, the Architect/Supervising Officer or the Employer or by any person or persons for whom they may be responsible or by any other person for whom the Nominated Supplier is not responsible;

·3 that delivery of the materials or goods supplied shall be commenced, carried out and completed in accordance with any delivery programme agreed between the Contractor and the Nominated Supplier or in the absence of such programme in accordance with the reasonable directions of the Contractor;

·4 that the Nominated Supplier shall allow the Contractor a discount for cash of 5 per cent on all payments if the Contractor makes payment in full within 30 days of the end of the month during which delivery is made;

·5 that the Nominated Supplier shall not be obliged to make any delivery of materials or goods (except any which may have been paid for in full less only any discount for cash) after the determination (for any reason) of the Contractor's employment under this Contract;

·6 that full discharge by the Contractor in respect of payments for materials or goods supplied by the Nominated Supplier shall be effected within 30 days of the end of the month during which delivery is made less only a discount for cash of 5 per cent if so paid;

·7 that the ownership of materials or goods shall pass to the Contractor upon delivery by the Nominated Supplier to or to the order of the Contractor, whether or not payment has been made in full;

·8 that if any dispute or difference between the Contractor and Nominated Supplier raises issues which are substantially the same as or connected with issues raised in a related dispute between the Employer and the Contractor under this Contract then, where * articles 5·1·4 and 5·1·5 apply, such dispute or difference shall be referred to the Arbitrator appointed or to be appointed pursuant to article 5 who shall have power to make such directions and all necessary awards in the same way as if the procedure of the High Court as to joining one or more defendants or joining co-defendants or third parties was available to the parties and to him and in any case the award of such Arbitrator shall be final and binding on the parties:

·9 that no provision in the contract of sale shall override modify or affect in any way whatsoever the provisions in the contract of sale which are included therein to give effect to clauses 36·4·1 to 36·4·9 inclusive.

36·5 ·1 Subject to clauses 36·5·2 and 36·5·3, where the said contract of sale between the Contractor and the Nominated Supplier in any way restricts, limits or excludes the liability of the Nominated Supplier to the Contractor in respect of materials or goods supplied or to be supplied, and the Architect/Supervising Officer has specifically approved in writing the said restrictions, limitations or exclusions, the liability of the Contractor to the Employer in respect of the said materials or goods shall be restricted, limited or excluded to the same extent.

·2 The Contractor shall not be obliged to enter into a contract with the Nominated Supplier until the Architect/Supervising Officer has specifically approved in writing the said restrictions, limitations or exclusions.

·3 Nothing in clause 36·5 shall be construed as enabling the Architect/Supervising Officer to nominate a supplier otherwise than in accordance with the provisions stated in clause 36·4.

*The Architect/Supervising Officer should state whether in the Appendix to the SFBC the words "Articles 5·1·4 and 5·1·5" have been struck out; if so then clause 36·4·8 will not apply to the Nominated Supplier.

Page 4

Fig 3.12 Architect's Instruction, from architect to contractor: removal of materials or goods not in accordance with the contract

```
[38] Remove from site the multi-colour sandfaced clay
     tiles supplied for tile hanging as these do not
     conform to the specification which calls for
     approved red colour.
```

Clause 8.4 of the Standard Form of Building Contract gives the architect power to order the removal from the site of any work, materials or goods not in accordance with the contract.
The method of notifying the contractor is by Architect's Instruction (see Fig 3.12). The instruction must carefully define the area of work or the precise material to which exception is made, and must state the reason for requiring its removal. This use of the Architect's Instruction ensures that the quantity surveyor, and clerk of works if employed, are automatically informed; the quantity surveyor will then omit any allowance for the offending item in the next interim valuation.

The IFC (Clause 3.14) gives the same authority for the removal of any works, materials or goods not in accordance with the contract.

The Agreement for Minor Building Works does not include any reference to this matter.

The contractor must comply with, and give all notices required by, any regulation or by-law of any local authority, or any statutory undertaking which has any jurisdiction over the works. In addition he must similarly comply with any Act of Parliament.
Under Clause 6.1.2 of the Standard Form of Building Contract, before any work is carried out to comply with these regulations or enforcements, including any directions given by a building control officer, the contractor should notify the architect in writing of the direction, giving the reason for the proposed variation. The architect should then issue an instruction confirming the work within seven days; if he does not do so the contractor is empowered to proceed with the work despite the fact that no instruction has been issued.

Warranty TNS/2

Schedule 3: Warranty by a Nominated Supplier

To the Employer:
named in our Tender dated

For:
(abbreviated description of goods/materials)

To be supplied to:
(job title)

1 Subject to the conditions stated in the above mentioned Tender (that no provision in this Warranty Agreement shall take effect unless and until the instruction nominating us, the order of the Main Contractor accepting the Tender and a copy of this Warranty Agreement signed by the Employer have been received by us) WE WARRANT in consideration of our being nominated in respect of the supply of the goods and/or materials to be supplied by us as a Nominated Supplier under the Standard Form of Building Contract referred to in the Tender and in accordance with the description, quantity and quality of the materials or goods and with the other terms and details set out in the Tender ('the supply') that:

1·1 We have exercised and will exercise all reasonable skill and care in:

1·1 ·1 the design of the supply insofar as the supply has been or will be designed by us; and

·2 the selection of materials and goods for the supply insofar as such supply has been or will be selected by us; and

·3 the satisfaction of any performance specification or requirement insofar as such performance specification or requirement is included or referred to in the Tender as part of the description of the supply.

1·2 We will:

1·2 ·1 save insofar as we are delayed by:

·1·1 force majeure; or

·1·2 civil commotion, local combination of workmen, strike or lock-out; or

·1·3 any instruction of the Architect/Supervising Officer under SFBC clause 13·2 (Variations) or clause 13·3 (provisional sums); or

·1·4 failure of the Architect/Supervising Officer to supply to us within due time any necessary information for which we have specifically applied in writing on a date which was neither unreasonably distant from nor unreasonably close to the date on which it was necessary for us to receive the same

so supply the Architect/Supervising Officer with such information as the Architect/Supervising Officer may reasonably require; and

·2 so supply the Contractor with such information as the Contractor may reasonably require in accordance with the arrangements in our contract of sale with the Contractor; and

·3 so commence and complete delivery of the supply in accordance with the arrangements in our contract of sale with the Contractor

that the Contractor shall not become entitled to an extension of time under SFBC clauses 25·4·6 or 25·4·7 of the Main Contract Conditions nor become entitled to be paid for direct loss and/or expense ascertained under SFBC clause 26·1 for the matters referred to in clause 26·2·1 of the Main Contract Conditions; and we will indemnify you to the extent but not further or otherwise that the Architect/Supervising Officer is obliged to give an extension of time so that the Employer is unable to recover damages under the Main Contractor for delays in completion, and/or pay an amount in respect of direct loss and/or expense as aforesaid because of any failure by us under clause 1·2·1 or 1·2·2 hereof.

Pages 1-4 comprising TNS/1 with Schedules 1 and 2 are issued in a separate pad.

Warranty TNS/2

2 We have noted the amount of the liquidated and ascertained damages under the Main Contract, as stated in TNS Schedule 1, item 8.

3 Nothing in the Tender is intended to or shall exclude or limit our liability for breach of the warranties set out above.

4·1 If at any time any dispute or difference shall arise between the Employer or the Architect/Supervising Officer on his behalf and ourselves as to the construction of this Agreement or as to any matter or thing of whatsoever nature arising out of this Agreement or in connection therewith; then such dispute shall be and is hereby referred to the arbitration and final decision of a person to be agreed between the parties hereto, or, failing agreement within 14 days after either party has given to the other a written request to concur in the appointment of an arbitrator, a person to be appointed on the request of either party by the President or a Vice-President for the time being of the Royal Institute of British Architects.

4·2 ·1 Provided that if the dispute or difference to be referred to arbitration under this Agreement raises issues which are substantially the same as or connected with issues raised in a related dispute between the Employer and the Contractor under the Main Contract or between a Nominated Sub-Contractor and the Contractor under Agreement NSC/4 or NSC/4a or between the Employer and any Nominated Sub-Contractor under Agreement NSC/2 or NSC/2a, or between the Employer and any other Nominated Supplier, and if the related dispute has also been referred for determination to an arbitrator, the Employer and ourselves hereby agree that the dispute or difference under this Agreement shall be referred to the arbitrator appointed to determine the related dispute; and such arbitrator shall have power to make such directions and all necessary awards in the same way as if the procedure of the High Court as to joining one or more of the defendants or joining co-defendants or third parties was available to the parties and to him.

·2 Save that the Employer or ourselves may require the dispute or difference under this Agreement to be referred to a different arbitrator (to be appointed under this Agreement) if either of us reasonably considers that the arbitrator appointed to determine the related dispute is not properly qualified to determine the dispute or difference under this Agreement.

4·3 Paragraphs 4·2·1 and 4·2·2 hereof shall apply unless in the Appendix to the Main Contract Conditions the words 'Articles 5·1·4 and 5·1·5 apply' have been deleted.

4·4 The award of such arbitrator shall be final and binding on the parties.

Signature of Supplier: ______________________________

Signature of Employer: ______________________________

Page 6

If, in an emergency, the contractor is required to provide materials or carry out work to comply with statutory requirements before receiving instructions from the architect, he must immediately inform him of the emergency and the action he is taking. So long as the reason is due to a divergence between the statutory requirements and all or any of the documents referred to in Clause 2.3, the work is deemed to have been carried out under an Architect's Instruction issued in accordance with Clause 13.2.

Clause 5.2 of the IFC is similar to the requirements of the SFBC Clause 6.1.2 and the extent of the contractor's liability is defined in Clause 5.3.

In an emergency the contractor is empowered under Clause 5.4 *et seq.* to carry out such limited work as to comply with statutory requirements and after informing the architect the contractor is entitled to receive Instructions dealing with the matter as a variation under Clause 3.6.

Clause 5.1 of the Agreement for Minor Building Works requires the contractor to comply with all notices required by statutory regulations or by-laws applicable to the works, and to pay all fees and charges legally recoverable (see Fig 3.13). Such charges are those levied by statutory authorities for service connections, and by local authorities for rating temporary buildings and hoardings. This clause does not, however, give any direction as to the issue of instructions on the matter.

Fig 3.13 Architect's Instruction, from architect to contractor: statutory obligations

[42] Increase the depth of the service duct rising in the corner of the kitchen by 100 mm to allow the rising main to be positioned 450 mm from the external wall.

This instruction confirms the requirement of the Southern Water Board set out in your letter dated 27 March 1976.

Interim payments to the contractor are a feature of most contracts. The usual period stipulated is monthly and while the certificate can be in any form, it is usual, and convenient where the Standard Form of Building Contract is concerned, to use that form prepared by the RIBA (see p. 95).
The certificate is printed in quadruplicate with a colour coding to facilitate distribution as follows:

The original certificate (no colour code).
Contractor's copy (yellow).
Quantity surveyor's copy (blue).
Architect's copy (magenta).

All are identical to facilitate carbon copying.
The original certificate (uncoloured) is sent to the employer with copies to the contractor, quantity surveyor and architect's file.
The contractor is entitled to payment within fourteen days of the presentation of the certificate to the employer. The amount indicated is the total value of work executed, plus the value of materials either on site or accepted by the architect for payment up to seven days prior to the issue of the certificate, less any monies previously certified and any retention. No payment may be certified for works not properly carried out. Monies may be certified: for special materials or equipment; for components stored in approved premises to ensure a degree of protection which cannot be provided on site; or for materials which are ready for fixing before the site is ready to receive them. The architect's discretion may be used here.
Retention monies usually amount to 5 per cent of the value of the works, unless a different percentage has been included in the contract.
Certificates are used to release retention monies as follows:

- Final payment to sub-contractor prior to final payment to contractor (subject to indemnification for latent defects).
- On the issue of a Certificate of Practical Completion. One moiety of the retention monies is released subject to certain provisions.
- On the employer taking possession of a portion of the premises. One moiety of the retention monies, related to the percentage value of the whole to the part, is released.
- On the issue of the architect's Certificate of Making Good Defects the residue of the retention monies is released, either for the part or whole of the premises as relevant.

Under the IFC (Clause 4.2 *et seq.*) interim payments are made in a similar manner to the Standard Form of Building Contract except that the manner of valuation is somewhat different (Clauses 4.2.1 and 4.2.2).

Interim certificate

and Direction

Architect:
address:

Employer:
address:

Contractor:
address:

Works:
situated at:

Job reference:

Interim Certificate No:

*Issue date:

Valuation date:

* This date is to be not later than 7 days from the Valuation date.

Original to Employer

Under the terms of the Contract dated

in the sum of £ for the Works named and situated as stated above

I/We certify that the following interim payment is due from the Employer to the Contractor; and

I/We direct the Contractor that the amounts of interim or final payments to Nominated Sub-Contractors included in this Certificate and listed on the attached *Statement of Retention and of Nominated Sub-Contractors' Values* are due to be discharged to those named.

Gross valuation inclusive of the value of Works by Nominated Sub-Contractors . . . . £

Less Retention which may be retained by the Employer as detailed on the Statement of Retention . . . . £

£

Less total amount stated as due in Interim Certificates previously issued up to and including Interim Certificate No. . . . . £

Amount due for payment on this Certificate . . . . £

(in words)

All the above amounts are exclusive of VAT

Signed Architect

Contractor's provisional assessment of total amounts included in above certificate on which VAT will be chargeable £ @ %

This is not a Tax Invoice

On practical completion, the architect shall within 14 days certify payment to the contractor of 97½ per cent of the total value of the works as defined in Clause 4.3 (see p. 97).

The issuing of a certificate in respect of the Agreement for Minor Building Works differs somewhat from the requirements of the Standard Form of Building Contract described above.
Interim certificates (described as 'progress payments') under Clause 4.2 are only made if specifically included in the contract on the contractor's request, otherwise Clause 4.2 is deleted and Clause 4.3 refers.
Under Clause 4.2 the architect certifies progress payments at four weekly intervals deducting a retention of 5 per cent (or such percentage as is applicable). The amount certified must be paid in full by the employer within fourteen days.
Under Clause 4.3 the architect is required to certify 97½ per cent of the value of work carried out within fourteen days of the date of practical completion, payment also being required from the employer in full within fourteen days.
The Certificate of Interim/Progress Payment is used for this form of contract, by omitting the references to Nominated Sub-contractors as the Agreement for Minor Building Works does not make provision for their employment.
Value Added Tax on supplies of goods and services was introduced under the Finance Act 1972 and came into force on 1 April 1973. From the provisions of this Act the following categories of supply by the contractor to the employer are generally applicable:

- New Works zero rated for VAT purposes.
- Works of repair, alteration or maintenance including goods and materials for the works. Charged at standard rate of VAT.
- The supply of materials or builders' hardware, sanitary ware or other articles of a kind *not* ordinarily installed by builders as fixtures. Charged at standard rate of VAT.
- Any supply not falling specifically within the three categories above. Charged at standard rate of VAT.

VAT payments are covered in the Standard Form of Building Contract under Supplemental Provision (the VAT Agreement) in the Intermediate Form of Contract under Supplemental Conditions A, and in the Agreement for Minor Building Works under Clause 5.2.
As these contract forms include provision for periodic payments each payment must be accompanied by any tax due, as this cannot wait the conclusion of the building contract. These contract forms are VAT exclusive. Employers who cannot claim credit for such tax or for only a

Certificate of

Interim/ Progress payment

Issued by:
address:

Employer:
address:

Serial no: A 113401

Contractor:
address:

Job reference:

Issue date:

Works:
Situated at:

Valuation date:

Contract dated:

Original to Employer

This certificate for interim/progress payment is issued under the terms of the above mentioned Contract in the sum of £__________

	A: Value of work executed and of materials and goods on site (excluding items included in B below)	£
1	Amount payable at_____%	£
2	___	£
3	B: Amounts payable (or deductible) in accordance with IFC 84 clause 4·2·2 at 100%	£
	Sub-total	£
	Less amounts previously certified	£
	Amount for payment on this Certificate	£

[1] Percentage is normally 95% except where Practical Completion has been achieved (97½%) or where some other percentage has been agreed by the parties.

[2] Space has been left for special payments such as at 'partial possession' and for goods and materials off site.

[3] This item applies only to IFC 84 Conditions and should be deleted where the Contract is MW 80. See Notes on the use of this form.

I/We certify that the amount for payment by the Employer to the Contractor on this Certificate is (in words)

____

____

All amounts are exclusive of VAT

To be signed by or for the issuer named above.

Signed__________

Contractor's provisional assessment of total amounts included in above certificate on which VAT will be chargeable £__________ @ _____%

This is not a Tax Invoice

Authenticated receipt

Original for Employer

Reference No:

Value Added Tax (General) Regulations 1980: Regulation 8(4)
Receipt by a Contractor supplying services, or services together with goods, to an Employer by way of sale under the Standard Form of Building Contract or other contract which provides for periodical payment without discount for such supplies.

This Receipt can be used by an Employer in place of a tax invoice to substantiate any input tax claimed.

* Contractor (the Supplier) ______________________

Contractor's address ______________________

____

Contractor's VAT Registration No. ______________________

* Employer (the person supplied) ______________________

Employer's address ______________________

____

* Contract Works ______________________

____

____

* *Insert the same names and descriptions as entered in the Articles of Agreement (Standard Form) or in any other main contract.*

1. Identifying number of Receipt ______________________

2. Date of the Supply ______________________

3. Description sufficient to identify the quantity of goods or the extent of the services and the amount, **excluding tax**, payable for each description.

		Value (tax-exclusive)
zero-rated		£
positively-rated at	%	£
at	%	£

4. Gross total amount payable excluding tax £ ______________________

5. Rate and amount of tax chargeable

Rate ______ Amount: £ ______________________

Received from the Employer the amount of tax of: £ ______________________

Signed: ______________________ Date ______________ 19

Contractor

16 PN 6/80

Appendix D

Contractor's authenticated receipt: Notes for guidance

General
Regulation 8(4) of the Value Added Tax (General) Regulations 1980 provides: 'Where the supplier of a supply to which Regulation 21 relates gives an authenticated receipt containing the particulars required under Regulation 9(1) to be contained in a tax invoice in respect of it, that document shall be treated as a tax invoice required to be provided under paragraph 1 of this regulation on condition that no tax invoice or similar document which was intended to be or could be construed as being a tax invoice for the supply to which the receipt relates is issued.'

The Receipts in this pad contain all the details which Regulation 9(1) requires to be included in a tax invoice. Under Regulation 9(1) the type of supply has to be stated and the Regulation sets out various descriptions. The Commissioners of Customs and Excise have confirmed that the one appropriate to supplies under the Standard Form is 'sale' so the supply is described as being 'by way of sale'. The intention is to differentiate the transaction from hirings, hire purchase and so forth.

The Regulation also provides that cash discount arrangements must be stated; since generally no cash discount is offered to the Employer the words 'without discount' are used. **If discount is offered the amount must be stated and 'value' is always the net amount after discount has been deducted whether or not the discount is earned.**

The Receipts are suitable for payments under the VAT provisions of the Standard Form of Building Contract and also for any other Form of Main Contract which provides for periodical payments.

The Receipts are printed in duplicate so that the Contractor can retain a copy which establishes the output tax due from him. Carbon paper is **not** required. To avoid damaging blank receipts the protector board must be inserted when completing a receipt manually.

Item 1
The identifying number can best be obtained by linking the number of the Receipt to the relevant Certificate Number for the Contract Works. Example: 'Certificate No. 1; Tax Receipt No. 1'.

Item 2
Under the Value Added Tax (General) Regulations, Regulation 21 the tax liability (that is the time of supply, or tax point, when the Contractor is liable for VAT) is the date when payment of the certificate, or the positively rated proportion of the value included in that certificate, is received by the Contractor. The date when payment of the certificate referred to at item 1 is received must be inserted i.e. the date when the cash or a cheque is received by the Contractor.

Item 3
The value of all the goods and services supplied in the relevant period and certified is to be referred to, not merely those which are positively rated. No further description is needed. The tax chargeable shown at item 5, is calculated only on the tax-exclusive positively rated value shown here. The amount received, excluding tax, must be divided between the zero-rated and positively rated supplies for which the amount has been paid. If the whole amount is for zero-rated supplies insert NIL under 'VALUE' opposite the first percentage for positively rated supplies. The additional percentage is provided if in future more than one rate of tax is in force.

Item 4
Always insert the cash received from the Employer under the Certificate No. inserted at item 1, that is **excluding** the tax. Any deduction that may have been made for liquidated damages must however be ignored and the actual cash received plus the liquidated damages is the amount to be inserted here.

Item 5
As the time of supply or tax point for the Contractor is when he receives the cash for which the Certificate has been issued, it is possible that there may be a change in rate etc. between the date when, under clause 1·1 of the VAT Agreement in the Standard Form or similar provisions in other Forms, the Employer calculates the tax payable under clause 1·1 and the date when the Contractor receives the VAT-exclusive payment of the relevant certificate. There is a possibility therefore that the Contractor may receive and issue the Receipt for an amount different from the amount that has to be inserted by him opposite item 5. This is the only reason why the amount of tax for which the receipt is issued will differ from the amount entered at item 5.

If the Contractor does not receive the amount for which he made an assessment of the value of his taxable supplies under clause 1·1 of the VAT Agreement clause 1·1 in the Standard Form (or equivalent provisions in other Forms) together with the tax calculated by reference to that value, he should regard the amount received as tax-inclusive. He should not therefore in these circumstances insert the tax he should have received at item 5 and the tax-exclusive amount he should have received at item 4; but instead he should, on a tax rate of 15%, regard three twenty-thirds of the amount received as the tax paid and insert that figure at item 5 and the balance of twenty twenty-thirds at item 4. If any other positive rate of tax applies a similar principle must be followed. (See Customs & Excise Notice No. 700 – 'VAT General Guide': Appendix D.)

PN 6/80 15

limited portion have rights of challenge for any VAT claimed from them by the contractor. To ensure that employers have some indication of the amounts likely to be charged in addition to the VAT-exclusive tender figure, the contractor may be required to state on his tender form the estimated tax liability in respect of the works. This estimate is in approximate form and intended only as a guide to liability.

Provisional VAT payments under the Standard Form of Building Contract and the Agreement for Minor Building Works made during the progress of the works are computed as follows:

1. The contractor, not later than the date for the issue of each Interim Certificate, shall give the employer a written provisional assessment of the value of goods and services against the value of the certificate and the rate of tax applicable.
2. The employer, on receipt of the assessment, calculates the amount of tax due and pays the amount together with the sum due under the certificate issued by the architect within the period (fourteen days) laid down in the contract for discharge of certificates.
3. If the employer has grounds for reasonable objection to the provisional assessment he must notify the contractor in writing within three days of receipt of the assessment.
4. After the issue of the Certificate of Completion of Making Good Defects under Clause 17.4 the contractor is required to issue a final statement showing the total value of all supplies for which certificates have or will be issued chargeable to him and the rates of tax applicable and issue this to the employer.

If the employer is taxable under VAT regulations he may claim input credit for VAT, paid against the authenticated receipt for the tax supplied by the contractor in the form issued by the Joint Contracts Tribunal (see pp. 98–99). If a receipt is not received by the employer for tax paid, he is under no obligation to make any further payment so long as he can show that he requires such a receipt to validate any claim on his own part for tax paid in respect of the contract works.

If the employer finds that the total amount of tax specified in the final statement exceeds the amount already paid by him, he must notify the contractor in writing of the overpayment and should receive a refund. This should be accompanied by a receipt showing correction of the amounts. Liquidated and ascertained damages applicable to the contract are exempted from the provisions of VAT.

Where sums are included in the architect's Interim or Final Certificate for payment to a nominated sub-contractor, the architect must direct the contractor as to the amount of such sums and the name of the

Statement of retention and of Nominated Sub-Contractors' values

Architect:
address:

Works:
situated at:

Relating to Valuation No:
Job reference:
Issue date:

	Gross valuation	Amount subject to:			Amount of retention	Net valuation	Previously certified	Balance due
		Full retention of %	Half retention of %	Nil retention				
	£	£	£	£	£	£	£	£
Main Contractor:								
Nominated Sub-Contractors:								
Total (The sums stated are exclusive of VAT)								

No account has been taken of any discounts for cash to which the Contractor may be entitled if discharging the balance within 17 days of the issue of the Architect's Certificate.

Architect.
address:

Employer:
address:

Contractor:
address:

Works:
situated at:

Nominated
Sub-Contractor:
address:

Notification

to Nominated
Sub-Contractor
concerning amount
included in
certificates

Job reference:
Interim Certificate No:
Issue date:
Valuation date:

Original to Nominated
Sub-Contractor

Under the terms of the Contract dated

I/We inform you that the amount of an
interim/final* payment due to you has been included in
Interim Certificate No. dated
issued to the Employer, in accordance with the attached *Statement of Retention and of Nominated Sub-Contractors' Values* and that the Contractor, named above, has been directed in the said Certificate to discharge his obligation to pay this amount in accordance with the terms of the Contract and the relevant Sub-Contract.

*Delete as appropriate

To comply with your obligation to provide written proof of discharge of the certified amount, you should return the acknowledgement slip below to the Contractor immediately upon such discharge.

Signed ______________________ Architect

Contractor:
address:

Works:
situated at:

Nominated Sub-Contractor's acknowledgement of discharge of payment due

Job reference:
Notification date:
Interim Certificate No:

We confirm that we have received from you discharge of the amount
included in Certificate No. ________ dated ________
as stated in the Notification dated ________
in accordance with the terms of the relevant Sub-Contract.

Signed ______________________ Nominated Sub-Contractor

Date ________

sub-contractor to which they are due (see p. 101). This is carried out by the issue of a Statement of Retention and of Nominated Sub-contractor's Values.
The statement is therefore issued with the certificate and completed as necessary.
The contractor is required to pay the sum due within seventeen days of the receipt of the architect's certificate and statement.

This procedure is only applicable to the Standard Form of Building Contract, under Clause 27. The IFC and Agreement for Minor Building Works make no requirements in this matter.

With the issue of the certificate and the statement, the architect should issue a Notification to a Nominated Sub-Contractor concerning the amount included in the certificate. The sub-contractor is then aware of the issue of the certificate and the date on which he could expect to receive payment, and on receipt of payment should complete the acknowledgement of discharge of payment due, forming a tear-off portion of the notification, and return it to the contractor (see p. 102).

Under Clause 6 of the Standard Form of Building Contract the architect is empowered to order the structure to be opened up:

- To confirm that thc work has bccn carried out as required and described in the contract.
- To investigate any defects which appear.
- To carry out any tests necessary on any executed work or materials.

Instructions to open up or expose the structure are given in an Architect's Instruction (see Fig 3.14).
If the work is found to be in accordance with the contract, or if the defect falls outside the contractor's liability, the architect will issue a further Instruction to make good and reinstate the work and direct that the cost both of opening up and reinstatement are to be charged as an extra to the contract.
If, however, the work or material is found to be defective or not in accordance with the contract, the architect will issue an Instruction to make good and reinstate the work, and direct that both opening up and reinstatement are to be at the contractor's sole cost.
The use of an Instruction ensures that automatic notification is given to both quantity surveyor and, if employed, clerk of works.

Fig 3.14 Architect's Instruction, from architect to contractor: open up works for inspection or testing

[46] Open up panelling to the main entrance where stained to allow inspection. Please advise architect when this work has been carried out.

Under the IFC (Clause 3.12) similar powers are given to the architect to open up and/or test materials/goods or work. The following Clauses 3.13.1 and 3.13.2 deal with the issue of Instructions following the failure of work etc.

The Agreement for Minor Building Works does not include any reference to this matter.

Under the Standard Form of Building Contract insurance against fire is covered by Clause 22 as follows:

- *Clause 22.A* refers to the insurance by the contractor in the joint names of contractor and employer against loss by specific perils.
- *Clause 22.A.4* requires the contractor on acceptance of any claim to restore, replace or repair any damaged unfixed materials or goods and remove and dispose of any debris. Any payment for such work plus the restoration or replacement of works, materials or goods is by instalments under architect's certificates only.
- *Clause 22.B* refers to new buildings where the risk is taken by the employer. The removal of debris, the repair or replacement of goods and the restoration of damaged work is deemed to be a variation requiring an Architect's Instruction.
- *Clause 22.C* refers to alterations or extensions to an existing building, where, if the works are damaged by fire or any other of the specified perils, the employment of the contractor may within twenty-eight days of the occurrence be determined at the option of either party. The removal and disposal of debris may be carried out by the

Fig 3.15 Architect's Instruction, from architect to contractor: restoration and repair after fire

[52] Under Clause 22.C.2.3.2 of the Standard Form of Building Contract remove all debris resulting from the fire in the roof on 13 June 1976. Remove the ceiling to Room 4, complete and replace the ceiling and roof in new materials as restoration including re-decoration of Room 4.

issue of an Architect's Instruction. The reinstatement and making good of damage is deemed to be a variation requiring an Architect's Instruction (see Fig 3.15).

Under the Intermediate Form of Contract, Clause 6.3 perils are defined in Clause 8.3 and include fire, lightning, storm and flood etc, and cover is provided under the contract in respect of these by means of insurance as provided for in Clause 6.3:

- *Clause 6.3A.1* refers to insurance by the contractor as Clause 22A of the SFBC.
- *Clause 6.3A.4* requires the contractor to restore damaged work as Clause 22.A.4 of the SFBC.
- *Clause 6.3B.1* refers to new buildings where the employer takes the risk and under Clause 6.3B.2 insures the works, reinstatement by the contractor being undertaken as a variation requiring an Architect's Instruction.
- *Clause 6.3C.1* requires the Employer to insure the works.

If any loss or damage occurs under Clause 6.3 perils, the contractor must give immediate notice to the employer of the extent and nature of the damage. The contract may within 28 days of the occurrence, at the option of either party and by registered post or recorded delivery, be determined. The provisions of Clause 7.7 will then apply.

The Agreement for Minor Building Works in Clause 6.3 requires insurance for the works, property and contents in respect of fire as follows:

Fig 3.16 Architect's Instruction, from architect to contractor: postponement of work

[63] Postpone the commencement of work to the interior of the building due to commence on [*insert date*], until [*insert date*]. The employer is now unable to give possession until the later date.

- *Clause 6.3A* refers to insurance by the contractor in the joint names of employer and contractor for new works. On acceptance of any claim the contractor must replace the damaged work, dispose of debris and complete the works and insurance payments shall be made against architect's certificates at the intervals laid down in Clause 4.0.
- *Clause 6.3B* refers to insurance by the employer for existing works. If any loss or damage occurs the architect is required to issue Instructions for the reinstatement and making good of the damage in accordance with Clause 3.5.

An Architect's Instruction must be issued to postpone any work to be carried out under the contract (see Fig 3.16).

Clause 23.2 of the Standard Form of Building Contract deals with this. Any direct loss or expense incurred by the contractor as a result of postponement will be valued under Clause 25, and the costs included in the next applicable interim payment.

Clause 3.15 of the Intermediate Form of Contract empowers the architect to issue instructions to postpone any work to be carried out under the contract.

Clause 3.6 of the Agreement for Minor Building Works empowers the architect to change the order or period in which any contract works are carried out, any such instruction to be valued by the architect on a fair and reasonable basis.

The architect is required to issue in writing a fair and reasonable extension of time in respect of certain delays so long as:

- He is notified by the contractor in writing of the cause of the delay as soon as the delay becomes apparent.
- The contractor uses the best of his ability to minimize the delay and to proceed with the works.
- The delays are outside the control of the contractor.

The architect should immediately assess the likely period of delay and notify the contractor as soon as possible of any extension of time he may be willing to grant by the issue of a Notification of an Extension of Time. The cost implications of any extension should be ascertained as soon as possible and reported to the employer (see p. 108).
Clause 25 of the Standard Form of Building Contract covers extension of time and the architect's duties.
The contractor is precluded from granting an extension of time to a nominated sub-contractor without first obtaining written consent from the architect.
Where the circumstances include reference to a nominated sub-contractor the contractor must send a copy of his notification of the cause of the delay to the sub-contractor concerned, and the architect shall send a copy of the Notification of Revision to Completion Date (see p. 110) to every nominated sub-contractor employed.

Clause 2.2 of the Agreement for Minor Building Works deals with the extension of time for reasons beyond the contractor's control.
The standard form, printed for use with the Standard Form of Building Contract, may be used for both contracts. The remainder of the form is self-explanatory.

Clause 2.3 of the Intermediate Form of Contract gives the architect power on receipt of written notice of the cause of the delay from the contractor, to make a fair and reasonable extension of time for completion of the works. Causes of delay which can be considered reasonable include:

- *Clause 2.4.5* Compliance with Architect's Instructions.
- *Clause 2.4.6* Opening up works shown to be in accordance with the contract.
- *Clause 2.4.7* Late receipt of instructions etc. from the architect.
- *Clause 2.4.8* Delays caused by work not forming part of the contract.
- *Clause 2.4.9* Delay caused by employer to supply agreed materials and/or goods.

Notification of an

Extension of time

Issued by:
address:

Employer:
address:

Contractor:
address:

Works:
Situated at:

Contract dated:

Serial no:

Job reference:

Issue date:

Under the terms of the above mentioned Contract,

I/We give notice that the time for completion is extended beyond the Date for Completion stated in the Contract so as to expire on:

______________________________ 19______

To be signed by or for the issuer named above.

Signed______________________________

Distribution

Original to: ☐ Employer

Duplicate to: ☐ Contractor

Copies to: ☐ Quantity Surveyor ☐ Services Engineer ☐ Structural Engineer ☐ File

o *Clause 2.4.12* Failure of employer to give in due time access to works etc.

as well as exceptionally adverse weather conditions, loss or damage due to Clause 6.3 perils, strikes etc.

Non-completion of sub-contract works by a nominated sub-contractor within the period defined by the completion date inserted in the sub-contract or any extension granted by the contractor with the architect's consent, requires the architect to certify the period within which the sub-contract works ought reasonably to have been completed.
Clause 26 of the Standard Form of Building Contract covers this requirement. Clause 2.4.5 (Clause 3.3 Named Sub-Contractors) in the Intermediate Form of Contract deals with this problem. The Agreement for Minor Building Works does not include any reference to this matter.
Problems can occur when the contractor does not insist on a proper sub-contract with each nominated sub-contractor, and does not agree in writing a programme of work.
Send the original to the contractor with a copy to the sub-contractor concerned. Other copies go to parties as applicable (see Fig 3.17).
If separate agreements between employers and sub-contractors are in force on a contract or if the architect either suspects or is informed by a sub-contractor that the contractor is failing to discharge all sums previously certified for payment to a nominated sub-contractor, the architect may ask the contractor to establish that he has paid all amounts as directed. Clause 35.13 of the Standard Form of Building Contract gives the architect the necessary authority.
If the contractor is unable or refuses to comply with the architect's request, the architect should issue a certificate of non-compliance detailing those payments which are in arrears and the employer may pay these sums direct to the sub-contractor and deduct the same from any monies due to the contractor. The sums detailed should be net after adjustment of cash discount and retention.

The Intermediate Form of Contract and the Agreement for Minor Building Works make no direction in this matter.

Send the original to the contractor with a copy to the employer. Other copies go to parties as applicable (see Fig 3.18).

Architect: address:

Employer: address:

Contractor: address:

Works: situated at:

Notification of Revision to
Completion Date

Job reference:
Serial No:
Issue date:

Under the terms of the Contract dated

I/We give notice that the Completion Date previously fixed

as ______________ is hereby:

*Delete as appropriate

*fixed later than that previously fixed;
*fixed earlier than that previously fixed;
*confirmed;

and is now

† Statement (a) is for revisions made **prior** to Practical Completion and (b) for revisions **after** Practical Completion. Delete as appropriate.

(a) †by reason of the relevant events identified in the Contractor's notices, particulars and estimates, and/or instructions requiring as a variation omission of work, which are set out below/overleaf:

(b) †by reason of our review pursuant to clause 25.3.3

Signed ______________ Architect

Original to:
☐ Contractor

Copies to:
☐ Employer
☐ Quantity Surveyor
☐ Clerk of Works
☐ Structural Consultant
☐ Services Consultant
☐ Electrical Consultant

Nominated Sub-Contractors:
☐
☐
☐
☐
☐
☐ Site

Fig 3.17 Notice of Nominated Sub-contractor's Non-completion

Architect's name and address

Notice of Nominated Sub-contractor's Non-completion

Job title and no.

To Contractor

Serial no.

Date

In accordance with Clause 26.4.2 of the Standard Form of Building Contract, I/we certify the following nominated sub-contract work ought to have been completed by

Date [*including any authorized extension of time*]

Nominated sub-contractor

Sub-contract works

A duplicate copy of this Certificate is being sent to the nominated sub-contractor.

Signature Architect/Supervising officer

Original to Contractor

Copies to Employer

Nominated Sub-contractor

Quantity surveyor

Structural/Civil engineer

Services consultant

Clerk of works

Architect's file

Fig 3.18 Notice of Non-compliance

Architect's name
and address

Notice
of
Non-Compliance

Job title and no.
To Contractor

Date

I/we certify that having been requested by me/us, as provided by Clause 35.13.3 of the Standard Form of Building Contract, the contractor has failed to furnish me/us with reasonable proof that all the amounts stated as due in previous certificates in respect of the total value of the work, materials or goods executed or supplied by the following nominated sub-contractor(s) have been duly discharged. The employer may forthwith pay such amounts direct to the nominated sub-contractor(s) and deduct the same from any sums due to the Contractor.

Nominated sub-contractor	Certificate no.	Gross amount certified	Sums not discharged

Signature ______________________ Architect/Supervising officer

Original to Contractor

Copies to Employer
Quantity surveyor
Structural/Civil Engineer
Services consultant
Clerk of works
Architect's file

4 The contract II: determination

Determination of the contract by the contractor under Clause 28 of the Standard Form of Building Contract may be summed up as:

1 Failure by the employer to honour certificates or the interference with or obstruction of the issue of certificates.
2 Suspension of the works for the period incorporated in the Appendix for causes covered in Clause 28.1.3, usually
- three months in respect of Clause 22 perils
- one month for any other reason

3 By reason of the bankruptcy of the employer (Clause 28.1.4)

The contractor must give notice by registered post or recorded delivery to either the employer or the architect determining his employment under the contract.
Immediately on so doing the contractor must remove from the site his temporary buildings, plant, tools, equipment, goods and materials and assist his sub-contractors to do the same, except for goods and materials for the works already paid for which become the property of the employer. The contractor is entitled to be paid for:

1 The total value of works completed.
2 Uncompleted work valued as if it were an authorized variation.
3 Direct loss and/or expense as ascertained under clauses 26 and 34.3.
4 The cost of materials ordered for the work.
5 Reasonable cost of removal of buildings, plant etc.
6 Direct loss and/or damage to contractor and subcontractors caused by the determination.

There are no directions in the contract respecting the issue of a certificate in respect of any payments due to the contractor under Clause 28.

Determination of the contract by the contractor under Clause 7.5 of the Intermediate Form of Contract may be on the following grounds:

1 Failure by the employer to pay an amount properly due to the contractor under:

- Clause 4.2 (Interim Payments);
- Clause 4.3 (Interim Payments on Practical completion);
- Clause 4.6 (Final Certificate).

2 Obstruction or interference by the employer in the issue of a due certificate (Clause 7.5.2).

3 Suspension of the whole or substantially whole of the uncompleted works for a continuous period of one month (Clause 7.5.3) by reason of:

- Failure of architect to issue instruction under Clauses 1.4, 1.6 and 3.15.
- Failure of architect to issue instructions specifically requested in writing by the contractor.
- Delay in carrying out works not forming part of the contract (see also Clause 3.11).
- Failure by the employer to give in due time access to the works after receipt by architect of notice by contractor.

Under the Agreement for Minor Building Works determination by the contractor under clause 7.2, may be for the following reasons:

1 Failure by the employer to make progress payments within fourteen days of payment being due.

2 Obstruction of the works by the employer or failure to make the premises available for the works on the date for possession given under the contract.

3 Suspension of the works by the employer for a continuous period of at least one month.

4 Bankruptcy of the employer.

In connection with the first three reasons, the contractor is required to serve notice of the default by registered post or recorded delivery allowing the employer seven days to remedy the fault. If he does not do so, the contractor may serve notice of determination of the contract in a similar manner.

Immediately on determining the contract, the contractor must remove his temporary buildings, tools and plant etc, and is entitled to be paid for:

1 The value of the work begun and carried out under the contract and for materials for the works on site.

2 Reasonable cost of removal of his buildings, plant etc.

The architect may give notice to the contractor of default on four grounds defined in Clause 27.1 of the Standard Form of Building Contract as follows:

- o If the contractor without reasonable cause wholly suspends the works before completion.
- o If he fails to proceed regularly and diligently with the works.
- o If he refuses or persistently neglects to comply with a written notice to remove defective work or improper materials or goods, which neglect materially affects the works.
- o If he fails to comply with Clause 19 regarding assignment or sub-letting.

Establishment of default requires the architect to maintain a careful check on both progress and quality of the contract works. Immediately a defect is noticed or established an Architect's Instruction must be issued to remedy the fault. The persistent ignoring of instructions to remedy a default cannot be tolerated. The architect should discuss the situation both with the employer and with his solicitor to receive instruction before proceeding with the issue of a formal letter giving notice under Clause 27.1. This letter should be dispatched either by registered post or by recorded delivery (see Fig 4.1).
If the contractor fails to conform to the notice within fourteen days of receiving the letter, the employer may, within ten days after continuation or repetition of the default, serve notice either by registered post or by recorded delivery terminating the employment of the contractor (see Fig 4.2).
The letter forming the notice of determination is generally prepared by the architect, on behalf of the employer, for his signature.

Determination by the employer under the Intermediate Form of Contract (Clause 7.1 *et seq.*) falls under four heads:

1 Suspension of the works by the contractor before completion without reasonable cause.
2 Failure by the contractor to proceed continuously with the works.
3 Neglect by the contractor to comply with a written notice or instruction from the architect to remove defective work or materials not in accordance with the contract.
4 Failure by the contractor to comply with:
 - o Clause 3.2 (Sub-contracting);
 - o Clause 3.3 (Named persons);
 - o Clause 5.7 (Fair wages).

The architect must give notice by registered or recorded delivery of the

Fig 4.1 Letter from architect to contractor: notice specifying default

To be sent by registered post or recorded delivery

Dear Sir

In accordance with Clause 27.1 of the Standard Form of Contract, I hereby notify you that you are in default in respect of:

1 Your failure to act on Architect's Instruction Number 49.

2 Your failure to comply with Clause 19.

If this default continues for a period of fourteen days following your receipt of this notice, the employer may determine your employment as contractor under this contract.

Yours faithfully

default and if the contractor persists for a period of fourteen days after receipt of the notice or subsequently repeats such default, then the employer may give notice by registered or recorded delivery determining the contract.

In addition to the above, under Clause 7.2 the bankruptcy of the contractor automatically terminates the contract but provision is included in this clause for the contract works to continue by arrangement between the employer and the contractor's trustee in bankruptcy.

Under the Agreement for Minor Building Works the employer may determine the contract by notice by registered post or recorded delivery to the contractor (see Fig 4.2).

The grounds under which the employer may determine the contract under Clause 7.1 of the Agreement for Minor Building Works are:

1 Failure to proceed diligently with the works without reasonable cause.

2 Suspension of the works before completion.

3 Bankruptcy of the contractor *or* a composition with his creditors

Fig 4.2 Letter from employer to contractor: determination by employer

To be sent by registered post or recorded delivery

Dear Sir

[*a*] In accordance with Clause 27.1 of the contract dated [*insert date*] between your company and myself, I hereby give notice that the default defined in the architect's notice of default dated [*insert date*] has continued for a period exceeding fourteen days.

Consequently, from the service of this notice, I hereby determine your employment as contractor, without prejudice to any rights or remedies to which I am entitled.

With the service of this notice, the rights and duties of the employer and contractor shall be as defined in Clause 27.4.

[*Or*]

[*b*] In accordance with Clause 7.1 of the Agreement for Minor Building Works, I hereby notify you that you are in default in respect of your failure to proceed diligently with the works, without reasonable cause.

Consequently, from the service of this notice, I hereby determine your employment as contractor without prejudice to any rights or remedies to which I am entitled.

Yours faithfully

or the issue of a winding up order *or* resolution for a voluntary winding up passed *or* a receiver or manager appointed *or* possession taken on behalf of any creditor.

Under Clause 27.4.2.1 of the Standard Form of Building Contract and within fourteen days of determination the architect (or employer) may require the contractor to assign to the employer the benefits of any agreements entered into for the supply of material or goods or the execution of works (see Fig 4.3).
If, however, the contract has been determined under bankruptcy (Clause 27.2) this right of assignment is not applicable.
The position of all sub-contracts for materials, goods or works must be investigated and the employer advised of any benefits accruing to him by the assignment of the existing agreements.
The issue of this letter must be with the employer's consent. Copies must be sent to the employer and other relevant parties.

Clause 3.1 of the Intermediate Form requires written consent from the other party for the assignment of the contract.

There are no directions as to assignment contained in the IFC or the Agreement for Minor Building Works.

Under Clause 27.4.3 (c) of the Standard Form of Building Contract and within fourteen days of determination the architect (or the employer) may require the contractor to remove from site part or all of the temporary buildings, tools, equipment, plant, goods or materials belonging to or hired by him (see Fig 4.4).
The architect will have advised the employer of any materials on site useful to him in completing the contract works. A schedule of items to be removed should be sent to the contractor with the appropriate letter within fourteen days of the date of determination. Unfixed goods and materials on site, of which the value has been certified, are the property of the employer and must not be included in this schedule. Copies must be sent to the employer and other relevant parties.

Under Clause 7.4 of the Intermediate Form of Contract the contractor is required to:
- Give up possession of the site.
- On receipt of instructions from the architect remove all plant, goods and materials belonging to him.

The employer has the right to remove and sell any plant etc, left on site after due notice of clearance has been made, employ and pay others to

Fig 4.3 Letter from architect to contractor: assignment of agreements

Dear Sir

Further to the notice of determination by the employer dated [*insert date*] I hereby require you, in accordance with Clause 27.4.2.1 of the contract, immediately to assign to the employer, without payment, the benefit of the following agreements entered into by your company with the respective sub-contractors and suppliers:

Steam Heating Co, heating and plumbing installation
Weatherall Roofing Co, bitumen felt roofing

Yours faithfully

Fig 4.4 Letter from architect to contractor: removal of materials from site

Dear Sirs

Further to the notice of determination by the employer dated [*insert date*], I hereby require you to remove from site all temporary buildings, tools, equipment, plant, goods or materials either belonging to or hired to you in accordance with the attached schedule.

Failure to remove these by [*insert date*] will empower the employer to remove and sell any such property and hold the proceeds, less any costs involved, to your credit.

Yours faithfully

carry out and complete the works (using at his discretion any plant etc, left on site) and on completion of the works set out an account showing his completion costs set off against the amount paid to the contractor before determination, the difference being a debt accruing to one or other of the original parties to the contract.

There are no directions as to the removal of materials from site in the Agreement for Minor Building Works.

Under Clause 27.4.4 of the Standard Form of Building Contract the contractor is liable for direct loss or damage caused to the employer by determination of the contract.
On completion of the works the architect must verify the accounts to enable him to certify any expenses properly incurred by the employer together with any direct loss or damage caused to him by determination (see Fig 4.5). The following parties must be informed of all developments:

- The surety, if one is involved, whose consent must also be obtained.
- The trustee or liquidator if the contractor is bankrupt or in liquidation.

The approval of the surety must also be obtained for the appointment of a new contractor to complete the works and the following matters:

1. The preparation of the detailed account, which is drawn up by the architect to include all direct costs, professional fees and expenses at the end of the subsequent completion works.
2. The final settlement, which is calculated on the difference between the theoretical final account of the original contractor and the actual cost of the subsequent works.

Copies must be sent to the employer and other relevant parties.

Under Clauses 7.8.1 and 7.8.2 of the Intermediate Form of Contract the contract may be determined by either employer or contractor if the works are suspended for a period of three months by reason of:

- Force Majeure.
- Loss or damage caused by Clause 6.3 perils.
- Civil commotion.

Determination under these clauses shall generally be in accordance with Clause 7.7. Neither the Standard Form of Building Contract nor the Agreement for Minor Building Works incorporate any directions specific to this matter.

Fig 4.5 Letter from architect to contractor: certification of amount of expenses in completion after determination

Dear Sir

I hereby certify that, to the best of my knowledge, the following additional expenses and direct costs have been incurred by the employer resulting from the determination of your contract dated [*insert date*].

1 Additional cost of completing the building works [*insert sum*].
2 Additional professional fees [*insert sum*].
3 Other direct loss and damages suffered by the employer [*insert sum*].
4 Legal fees and expenses [*insert sum*].

Yours faithfully

5 The contract III: arbitration

In many business matters disagreements are bound to occur over timing, quality or delivery, to mention but simple instances. Building is no different from any other commercial enterprise and disputes between employer and contractor occur for a number of reasons. In view of its continuing nature, a building contract cannot be delayed for long periods while the law argues over the rights and wrongs of the matter. Consequently the Joint Contracts Tribunal have included in all building contracts issued by them an Arbitration Clause which not only defines matters that may be referred to arbitration but also lays down the procedure to be followed in referring these matters to an arbitrator.

Arbitration under all forms of contract is subject to the provisions of the Arbitration Acts 1950 and 1979. Under the Arbitration Act 1950, unless specifically requested by the parties, the arbitrator was not required to give any more than a bare decision on each matter in dispute. However, under the 1979 Act it is usual to give reasons and, if approached, the High Court may order the arbitrator to state his reasons, although these are not a statutory requirement. The court will not order reasons to be given unless it is satisfied that:

1 One or more parties, prior to the award being made, gave notice that a reasoned award was required and

2 There was some special reason why such notice was not given to the arbitrator.

Generally, it is unnecessary and not desirable to incorporate evidence in the reasonings: a summary of the background to the dispute, the issues between the parties and the arbitrator's findings and conclusions suffice.

The Arbitration Act, 1979 also incorporates other amendments to previous legislation:

1 The High Court no longer has the power to set aside or remit an award on grounds of error of fact or law on the face of the award. However, either party has a restricted right of appeal on points of law.

2 Any appeal to the Court of Appeal requires leave of the High Court or Court of Appeal and the legal question involved has to be of general importance or on some special point.
3 The parties may agree to exclude right of appeal after the arbitration has commenced to ensure that the dispute is settled by the arbitrator, whose decision and award is then final.

Arbitration agreements, including those in JCT contracts covered by the relevant arbitration clauses, allow for either party to refer a matter for arbitration. The submission must state the grounds on which arbitration is required and, failing agreement by both parties on the appointment of a specific person to act as arbitrator, the appointing body must be named but not necessarily the arbitrator. The submission must be precise as arbitration can only proceed on the items mentioned and these are limited to those specifically defined in the arbitration agreement.
The arbitrator always directs in his award as to who is to pay costs, unless specifically requested not to do so direct. Costs are at the discretion of the arbitrator and as a general rule 'costs follow the event'. Disputants can, however, agree between themselves as to the appointment of costs.
The Arbitration Clause in the JCT Standard Form of Building Contract is Article 5 and the method of its use is in accordance with the Arbitration Acts 1950 and 1979. Signature of both parties to the contract is deemed signature to an Agreement of Arbitration. Article 5 covers seven main items dealing with disputes which may arise:

1 On the construction of the contract.
2 On any matter arising from actions by the architect left to his discretion under the contract clauses.
3 Or the withholding by the architect of any certificate to which the contractor claims entitlement.
4 Or in respect of the adjustment of the contract sum under Clause 30.6.2 (Items included in adjustment of Contract Sum).
5 Or in respect of the rights and responsibilities of the parties under Clauses 27 and 28 (Determination).
6 Or in respect of the rights and responsibilities of the parties in respect of Clauses 32 (Outbreak of hostilities) and 33 (War damage).
7 Or in respect of the unreasonable withholding of consents or agreement by the employer or the architect acting on his behalf, or by the contractor.

Certain contractual matters are, however, expressly excluded from inclusion in matters that are covered by Article 5 namely:

1 Clause 31 (Finance (No. 2) Act 1975) as provided in Clause 31.9.

Fig 5.1 Letter from architect to RIBA: request for Form A relating to appointment of an arbitrator

The President
Royal Institute of British Architects
66 Portland Place
London
W1N 4AD

Dear Sir

With reference to the contract dated [*insert date*] and made between [*insert name*] of the one part and [*insert name*] of the other part, the contract form being the Standard Form of Building Contract [*Private Edition With/Without Quantities ... Edition ... Revision*] and where there is a submission to arbitration within the meaning of the Arbitration Acts 1950 and 1979 of a dispute between the employer (or architect acting on his behalf) and the contractor, I would be obliged if you would send to me a form a relating to the Appointment of an Arbitrator by the President of the Royal Institute of British Architects.

Yours faithfully

Signed I. Jones RIBA
for Line and Wash, architects

2 Clause 3 of the VAT Agreement (Employer's right to challenge tax claimed by contractor).
3 Clause 19A (Local Authorities Edition – Fair Wages Clause).

On submission by either party to the contract of particulars of the dispute to the other and in the event of no agreement being reached on the appointment of a suitable arbitrator a written request may be made on the request of either party to the President or a Vice-President for the time being of the Royal Institute of British Architects to nominate a person to act as arbitrator in the dispute (see Figs 5.1, 5.2, 5.3 and p. 127).
In most contracts the hearing takes place after the practical completion of the contract or termination of the contractor's employment. In some instances this may not be convenient as the contract may be held up until the dispute is resolved. In such cases an immediate arbitration can be undertaken on the following grounds:

1 Under Articles 3 or 4 dealing with the appointment of an architect or a quantity surveyor. Either of these may have ceased to act for some reason or other and the employer's nomination may not be acceptable to the contractor.
2 Under Clause 4.2 dealing with the issue of an Architect's Instruction.
3 Under Clause 30.1.3, a certificate is improperly withheld.
4 Under Clause 30.1.1.1, a certificate is not in accordance with the Conditions.
5 Any dispute or difference under:
- Clause 4.1 (compliance with Architect's Instruction)
- Clause 25 (Extension of time)
- Clause 32 (Outbreak of hostilities)
- Clause 33 (War damage)

While the powers of the arbitrator are wide he cannot exceed those of the architect granted under the contract and in this respect he is restricted to the provisions of the contract. He can, however, order measurements and valuations to determine the truth and open up and review certificates. He cannot question the powers of the architect to issue instructions under Clause 4.1 unless these powers are challenged by the contractor (Clause 4.2). The Final Certificate must be challenged before issue by the employer and within fourteen days of issue by the contractor. If arbitration proceedings have been commenced by either party before the issue of the Final Certificate, then the issue of such a certificate shall have effect as conclusive evidence as provided in Clause 30.9.1.

Royal Institute of British Architects 66 Portland Place London W1N 4AD

Appointment of an Arbitrator

Note: This form is only to be used where the dispute arises under a building contract wherein there is provision for arbitration.

Regarding the agreement dated the__________day of____________________ and made

between __________________________________of the one part

and__________________________________of the other part and where there is a submission to arbitration, within the meaning of the Arbitration Act 1950, of any dispute or difference which shall arise between the Employer, or the Architect acting on his behalf, and the Contractor.

This part only to be filled in by applicant(s)

WHEREAS a dispute or difference has arisen in connection therewith

I/We hereby ask the President of the Royal Institute of British Architects to nominate an arbitrator to hear and determine the matter.

As a Condition of the Appointment I/We jointly and severally agree as follows:

(i) To provide adequate security for the due payment of the fees and expenses of the Arbitrator if he so requires.

(ii) To pay the fees and expenses of the Arbitrator whether the Arbitration reaches a Hearing or not.

(iii) To take up the Award (if any) within ten days from receipt of notice of publication.

(Signed)__________________________________

of__________________________________

(Signed)__________________________________

of__________________________________

Dated this________________day of________________19____

I hereby appoint__________________________________

of__________________________________

Arbitrator in the above matter.

(Signed)__________________________________

President of the Royal Institute of British Architects

Dated this________________day of________________19____

I hereby accept the appointment of Arbitrator in the above matter.

(Signed)__________________________________

Dated this________________day of________________19____

Fig 5.2 Letter from architect to President, RIBA: covering return of Form A

The President
Royal Institute of British Architects
66 Portland Place
London W1N 4AD

Dear Sir

I return duly completed the form requesting the Appointment of an Arbitrator in accordance with:

Article 5 of the Standard Form of Building Contract

[*Or*]

Article 5 of the Intermediate Form of Contract

[*Or*]

Article 4 of the Agreement for Minor Building Works

and look forward to receiving your nomination in due course.

Yours faithfully

Signed I. Jones RIBA
for Line and Wash, architects

Fig 5.3 Letter from architect to contractor: asking for agreement to appointment of named arbitrator

A. B. Construction Co Ltd
49 High Street
New Town

Dear Sirs

With reference to the contract dated [*insert date*] and made between [*insert name*] of the one part and your company [*insert name*] of the other part and where in the matter of [*the issue of the Penultimate Certificate*] you have made formal request in writing for the matter to be referred to arbitration, I have to ask you if you accept the appointment of Mr Comyn A. Long RIBA as Arbitrator in this matter.

If you accept I would be obliged if you would indicate your agreement in writing so that I can formally invite Mr Long to accept nomination.

Yours faithfully

Signed I. Jones RIBA
for Messrs Line and Wash, architects

If arbitration proceedings have been commenced within fourteen days of the issue of the Final Certificate, this shall be conclusive evidence as provided in Clause 30.9.1 save in the matters to which the arbitration proceedings relate.
Generally, the Final Certificate can only be opened up for investigation in respect of:

1 Arithmetical errors or
2 As provided in Clause 30.9.1 save only in respect of all matters to which the proceedings relate.'

The award of the arbitrator is final and binding on all parties.

The arbitration clause (Article 5) in the Intermediate Form of Contract can generally be construed in a similar manner to the Standard Form of Building Contract but in Article 5.3 expressly states that when it is stated in the Appendix that this article shall apply that either party:

1 may appeal to the High Court to determine any point of law arising from the reference
2 may apply to the High Court to determine any question of law arising out of an award made under this Arbitration Agreement

and parties to the agreement accept the High Court's jurisdiction to determine such points of law. When the matters in dispute are substantially those that occur in a related dispute under a sub contract and where Article 5.4 is stated in the Appendix to apply and the related dispute has already been referred for determination by an arbitrator, then the related dispute is to be referred to the appointed arbitrator.
Three matters are expressly deleted from those on which either party to the Agreement may request arbitration (Article 5.1) and these are:

- Disputes under clause 5.7 (Fair Wages).
- Supplemental Condition A7 (Employer's right to challenge tax claimed by contractor).
- Supplemental Condition B8 (Application of Arbitration Agreement).

In addition certain restrictions are place on both parties by the operation of:

1 Clause 3.5.2, where the contractor must, after requesting the architect's authority for the issue of an instruction and – having received in writing from the architect details of his contractual authority – complies with the instruction, is thereby precluded from requesting the appointment of an arbitrator to decide on the architect's power to issue the instruction.
2 Clause 4.7, where either party must make a request for

arbitration within twenty-one days of the date of issue of the final certificate in respect of the proviso to Clause 1.1 (Contractors obligation).
3 Supplemental Condition D8 (Formulae Fluctuations).

Otherwise, under Article 5 the powers of the arbitrator to order measurements, valuations, open up certificates, opinions, decisions etc, are similar to those applicable under the Standard Form of Building Contract.

The arbitration clause in the Agreement for Minor Building Works is Article 4 and again its method of use is governed by the Arbitration Acts, 1950 and 1979. Signature by both parties to the agreement is signature to an Agreement of Arbitration. Article 4 covers three matters dealing with any disputes which may arise:

1 The arbitrator has powers to deal with disputes or differences between the employer (or the architect/supervising officer on his behalf) and the contractor on the construction or interpretation of the contract or the works.
2 The arbitrator may be a person agreed between the parties or, by failure to agree on a person within fourteen days after one of the parties has submitted a written request to agree to the appointment of an arbitrator, a person to be appointed on the request of either party by the President of the Royal Institute of British Architects.
3 No time is fixed for the hearing of the dispute. Timing is left to the disputants to settle as best to suit the particular circumstances of the dispute.

The main conditions for an arbitration are:

1 There must be a dispute.
2 There must be an agreement to refer the dispute to a nominated person.
3 The proceeding must be quasi-judicial with formal submissions and evidence.
4 The awards must be final, apart from certain matters of appeal, and are outside the jurisdiction of the Courts except where expressly noted to the contrary.

The main advantages of an arbitration are:

1 The employment of a specialist in building matters as arbitrator, usually an architect or chartered surveyor.
2 The decision or award is usually given more quickly than action at law.

3 Arbitration is generally very much less costly.

4 In most cases the parties can arrange for the date of the arbitration hearing to suit their convenience.

5 The arbitrator can inspect the works with little inconvenience.

6 The hearings are private and not subject to Press report.

The only disadvantage of arbitration compared with court proceedings is that there is no case history or record and each arbitration proceeds on its own evaluation. There are some moves being made to record, with the approval of the disputants, some aspects of arbitration awards but this is more a problem for the legal profession than either the arbitrator or parties to building contracts.
(For details of arbitration procedures and the duties of the Architect as Arbitrator reference is made to the publication of that title obtainable form RIBA Publications Ltd.)

6 The contract IV: the defects liability period

Practical completion of the works can be either 'complete' or 'partial'. With the consent of the contractor, under Clause 18 of the Standard Form of Building Contract, the employer may take possession of a portion of the works prior to possession of the whole. Total completion of contract works is rarely achieved by the completion date inserted in the contract. So long as the works are sufficiently completed to provide the employer with the accommodation and services necessary for his occupation, the practical completion can be certified. The completion of small items of work and minor corrections can usually be cleared during the employer's occupation.
The Certificate of Partial Possession by the employer (see p. 135) is issued by the architect when in his opinion practical completion of a portion of the works has occurred which the contractor has agreed may be taken possession of by the employer. This certificate must be issued within seven days of the date of partial possession and must incorporate an approximate value of the relevant part of the works. The effect of the issue of this certificate is:

1 The defects liability period in respect of the relevant portion of the works is deemed to commence on the date of partial possession.
2 The contractor is entitled to receive a certificate for the release of a moiety of the retention moneys attributable to the value of that part of the works covered and valued in the certificate.
3 The contractor is entitled to reduce his cover under Clause 22 perils by the value of the relevant works, which part is now at the sole risk of the employer.
4 Liquidated damages are reduced in respect of the balance of the works in the same ratio as the value of the relevant works to the whole contract.
5 The contractor is no longer liable for damage from frost to the relevant works.

When in the opinion of the architect practical completion of the whole

works has been reached he must issue a Certificate of Practical Completion of the Works (see p. 136). This form is dual purpose and allows for practical completion of a sub-contract (e.g. the heating and ventilation installation after acceptance by the services consultant) as well as for completion of the main contract. The effect of the issue is similar to that in respect of the partial completion. No date is inserted in the contract for the issue of this certificate.
The date inserted in both certificates for partial or practical completion signifies that for the period incorporated in the Appendix to the contract (Clause 17.2) any defects, shrinkages or other faults appearing within this period due to materials or workmanship not in accordance with the contract or to frost occurring before the relevant date are to be made good at the contractor's sole cost.

Under Clause 2.9 of the Intermediate Form of Contract the architect must issue a Certificate of Practical Completion when in his opinion the works are practically complete and the Defects Liability Period commences on the date named in the certificate (see p. 137).

Under Clause 2.4 of the Agreement for Minor Building Works, the architect is required to certify the date when in his opinion the works have been practically completed. This certificate can take the form of a letter and should certify the period during which the contractor is liable for defects as defined in Clause 2.5 (see Fig 6.1). A copy should be sent to the employer for his information. In addition, under Clause 4.3, when no previous certificates have been issued, the contractor is entitled to the issue of a certificate for 97½ per cent of the total amount due to him under the contract, including extra payment for work as agreed under Clauses 3.6 and 3.7. This certificate is due within fourteen days after the certified date of practical completion.

Under all three forms of contract, provision is made for liquidated and ascertained damages. In the Standard Form of Building Contract this is set out in Clause 24.2 and the sum allowable for each week is inscribed in the Appendix. In the Intermediate Form of Contract, Clause 2.7 requires that subject to the issue of a Certificate of Non-completion liquidated damages at the rate stated in the Appendix for the period of non-completion shall be deducted from monies due to the contractor. In the Agreement for Minor Building Works, Clause 2.3 deals with this matter and the sum allowable is inscribed in the blank space provided in the clause itself (see p. 138).

It is necessary to take care to maintain close control over likely causes of

Architect:
address:

Employer:
address:

Contractor:
address:

Works:
situated at:

Certificate of
Partial possession
by the Employer

Job reference:
Serial No:
Issue date:

Original to Employer

Under the terms of the Contract dated ______________

I/We certify that a part of the Works, referred to as the relevant part, namely:

the approximate value of which I/we estimate for the purposes of this Certificate, but for no other purpose, to be .... £ ______________

was taken into possession by the Employer on: ______________ (date)

and that for this relevant part of the Works only the Defects Liability Period will end on: ______________ (date)

Signed ______________________________ Architect

The Employer should note that as from the date of possession of the relevant part, he becomes solely responsible for the insurance of the relevant part of the Works.

Copies to:
__ Contractor
__ Quantity Surveyor
__ Clerk of works
__ Structural consultant
__ Services consultant
__ Electrical consultant
Nominated Sub-Contractors
__
__
__
__
__ File

Certificate of

Practical completion

of the Works
OR of works executed
by a Nominated
Sub-Contractor

Architect:
address:

Employer:
address:

Contractor:
address:

Works:
situated at:

Nominated
Sub-Contract
Works:
(if applicable)

Job reference: ____________

Serial No: ____________

Issue date: ____________

Under the terms of the Contract dated ____________

I/We certify that:

*Delete as appropriate

*1. Practical Completion of the Nominated Sub-Contract Works referred to above, was achieved on:

____

*2. Practical Completion of the Works was achieved on:

____

The Employer should note that as from the date of issue of this Certificate of Practical Completion of the Works the Employer becomes solely responsible for insurance of the Works.

Signed ____________ Architect

Original to:	Copies to:		Nominated Sub-Contractors:	
☐ Employer	☐ Contractor	☐ Structural Consultant	☐ ____________	☐ ____________
	☐ Quantity Surveyor	☐ Services Consultant	☐ ____________	☐ ____________
	☐ Clerk of Works	☐ Electrical Consultant	☐ ____________	☐ Site

Certificate of
Practical Completion
of the Works

Issued by:
address:

Employer:
address:

Contractor:
address:

Works:
Situated at:

Contract dated:

Serial no:

Job reference:

Issue date:

Under the terms of the above mentioned Contract,

I/We certify that Practical Completion of the Works was achieved on:

________________________________19________

To be signed by or for the issuer named above.

Signed____

The Defects Liability Period will therefore end on:

________________________________19________

Distribution	Original to:	Duplicate to:	Copies to:	
	☐ Employer	☐ Contractor	☐ Quantity Surveyor	☐ Services Engineer
		☐ Nominated Sub-Contractors	☐ Structural Engineer	☐ File

Certificate of

Non-completion

Issued by:
address:

Employer:
address:

Contractor:
address:

Works:
Situated at:

Contract dated:

Serial no:

Job reference:

Issue date:

Under the terms of the above mentioned Contract,

I/We certify that the Contractor has failed to complete the Works by the Date for Completion or within any extended time fixed under the contract provisions.

To be signed by or for the issuer named above.

Signed____________________

Distribution

Original to: ☐ Employer

Duplicate to: ☐ Contractor

Copies to: ☐ Quantity Surveyor ☐ Structural Engineer ☐ Services Engineer ☐ File

Fig 6.1 Letter from architect to contractor: certification of practical completion, Agreement for Minor Building Works

Dear Sir

As required under Clause 2.4 of the agreement dated [*insert date*] between your company and [*insert employer's name*] I certify that the contract works were practically completed on [*insert date*] and that defects liability as defined in Clause 2.5 will terminate on [*insert date usually three months subsequent to date of practical completion or period included in the contract*].

Yours faithfully

delay, and to agree as far as possible any revisions to the date for completion. Under both forms of contract the contractor is required to allow to the employer such sums as are due under liquidated and ascertained damages for agreed delay in completion of the works, as shown in Clause 24.2 of the Appendix to the contract, and the employer may deduct any such sums from monies due to the contractor.
The usual method of notifying both parties to the contract of delay in completion is the issue of a Certificate of Delay in Completion (see Fig 6.2). This is a simple form and when completed the original is sent to the contractor with copies to the appropriate parties including the employer.

The certificate shown in Fig 6.2 has been prepared for use in conjunction with the Standard Form of Building Contract. This can be varied for use with the other forms of contract as follows:

- Intermediate Form of Contract: Delete 24.1 and replace with 2.6. Delete 25 and replace with 2.3 and 2.5.
- Agreement for Minor Building Works: Delete 24.1 and replace with 2.1. Delete 25 and replace with 2.2.

If a defect or an emergency occurs under the Standard Form of Building Contract, which is due to omission, to materials or workmanship not being in accordance with the contract, to neglect by the contractor, or to

Fig 6.2 Notice of Delay in Completion

Architect's name
and address

Notice of
Delay in Completion

Job title and no.

To contractor

Serial no.

Date

In accordance with Clause 24.1 of the Standard Form of Building Contract, we certify that in our opinion, the works ought reasonably to have been completed on [*insert date*] after due allowances for any extended time fixed by us under Clause 25.

Signature

Architect/Supervising officer

Original to Contractor

Copies to Employer

Sub-contractor

Structural/Civil engineer

Quantity surveyor

Services consultant

Clerk of works

Fig 6.3 Architect's Instruction, from architect to contractor: to make good a defect within the Defects Liability Period

[27] Repair faulty valve on cold water down service and make good damage to decorations and floor.

This work is to be carried out forthwith and at your sole cost. Please agree date and time for access with the employer.

any other circumstance that is his liability, the architect may give immediate instructions for the defect to be made good.

The instructions are given in writing, usually in the form of an Architect's Instruction (see Fig 6.3). Any such defect is usually an emergency and the contractor should be told to deal with it as a matter of urgency and to make good any consequential damage at the same time. It is as well to give directions as to costing so that no ambiguities of this nature arise.

If, however, a defect occurs under the Standard Form of Building Contract after the delivery of a schedule of defects, no such instruction can be issued to deal with matters of this nature. Neither can instructions be issued after fourteen days subsequent to the expiry of the Defects Liability Period.

The Intermediate Form of Contract and the Agreement for Minor Building Works make no specific reference to these matters, merely requiring defects for which the contractor is liable to be made good within the stated period.

Any defects, shrinkage or other faults appearing within the Defects Liability Period and which are due to materials and workmanship not in accordance with the contract must be specified by the architect in writing (see Fig 6.4).

This usually takes the form of a Schedule of Defects and copies are sent to the contractor for his attention and the employer for his notification (see Fig 6.5). It is usual for the employer to draw the attention of the architect to items for inclusion in the list of defects. This list should be

Fig 6.4 Letter from architect to contractor: to enclose Schedule of Defects

Dear Sir

[*a*] In accordance with Clauses 15 (2) and 1(1) of the Standard Form of Building Contract dated [*insert date*] for the above works, I have carried out an inspection and attach herewith two copies of my Schedule of Defects for your immediate attention.

[*Or*]

[*b*] In accordance with Clause 2.10 of the Intermediate Form of Building Contract dated [*insert date*] for the above works, I have carried out an inspection and attach herewith two copies of my Schedule of Defects for your immediate attention.

[*Or*]

[*c*] In accordance with Clause 9(ii) of the Agreement for Minor Building Works dated [*insert date*] for the above works, I have carried out an inspection and attach herewith two copies of my Schedule of Defects for your immediate attention.

Will you please advise me immediately when you have completed the work so that I may carry out a further inspection.

Yours faithfully

Fig 6.5 Schedule of Defects

New Building
South Road
New Town

For Messrs A. Client & Co Ltd

Job number 259

Contractor: Messrs A. B. Construction Co Ltd,
49 High Street, New Town

1 Ease and adjust doors D.1 and D.17 binding on frames and make good paintwork.

2 Remake joint on WC pan/soil stack to staff cloaks (female), Room 27.

3 Clear away from site 2 no. precast concrete road kerbs and small quantity of ballast left on north side of car park.

4 Burn off defective paintwork to fascia, prime, stop and refinish as specified in contract.

[*Insert date*]

Signed: I. Jones RIBA
Messrs Line & Wash, architects
The Muse
Olympia

obtained before the architect carries out his inspection and those items which are relevant under the contract should be included in the Schedule of Defects.

Under the Standard Form of Building Contract Clause 17.2 the schedule must be delivered to the contractor within fourteen days of the expiry of the Defects Liability Period. The Intermediate Form of Contract in Clause 2.10 makes a similar condition. The Agreement for Minor Building Works makes no such stipulation but the schedule should be delivered to the contractor before the end of the Defects Liability Period to conform to the limitation contained in Clause 4.4.
All defects must be made good by the contractor at his sole cost within a reasonable time. Under the Standard Form of Building Contract the architect is empowered to instruct otherwise as to cost. Frost damage subsequent to practical completion is exempted from defects liability (Clause 17.5).

Any work extra to the contract ordered at the end of the Defects Liability Period should be kept entirely separate from the Schedule of Defects (see Fig 6.6). It may be possible to agree rates for the work but it is unlikely that the contractor will agree to the use of the contract rates if quantities form part of the contract. The best method is to ask for a separate lump estimate for the work and keep the whole separate from the original contract.

Fig 6.6 Letter from architect to contractor: works extra to contract

Dear Sir

The employer wishes to extend the garden wall and associated pavings for a further 6 m towards the rear of the building, all to match existing.

I enclose a drawing numbered [*insert number*] showing the proposals in detail and I should be pleased to have your firm lump sum estimate for the work within two weeks.

Yours faithfully

Architect:
address:

Employer:
address:

Contractor:
address:

Works:
situated at:

Certificate of completion of

Making good defects

Job reference:
Serial No:
Issue date:

Original to Employer

Under the terms of the Contract dated ____________

I/We hereby certify that the defects, shrinkages and other faults specified in the schedule of defects delivered to the Contractor as an instruction have in my/our opinion been made good.

This Certificate refers to:

*Delete as appropriate

*1. The Works described in the Certificate of Practical Completion
Serial No. ________ dated ____________

*2. The Works described in the Certificate of Partial Possession of a relevant part of the Works
Serial No. ________ dated ____________

Signed ______________________________ Architect

Date ____________

Copies to:
__ Contractor
__ Quantity Surveyor
__ Clerk of works
__ Structural consultant
__ Services consultant
__ Electrical consultant
Nominated Sub-Contractors
__
__
__
__
__ File

All three forms of contract require certification for the making good of defects, faults and shrinkages appearing during the Defects Liability Period. This is set out in the Standard Form of Building Contract under Clause 17.4 and 18.1, in Clause 2.10 of the Intermediate Form of Contract, and in the Agreement for Minor Building Works under Clause 2.5.
The standard form is applicable for all forms of contract, although specifically prepared for the Standard Form of Building Contract.
The form is prepared in two parts. Part 1 is for the making good of defects on contracts or contract works covered by Clause 17.1 (practical completion of whole works). Part 2 relates to the making good of defects on partial completion covered by Clause 18.1.1. The appropriate section shall be deleted.

As the Agreement for Minor Building Works makes no provision for partial completion, Part 2 does not apply and should be deleted. Part 1 should be amended by the deletion of 'Certificate of Practical Completion Serial No' and the insertion of 'letter' and the date of that certifying practical completion.
Immediately on issue of the Certificate of Making Good Defects (see p. 145), the remaining moiety of the retention monies must be released. This is obtained through an architect's certificate instructing the sum involved. This procedure is a requirement only of the Standard Form of Building Contract. The other two contracts make no provision for such release.

7 The contract V: the final account

The preparation of the statement of Final Account is generally carried out either:

- by the architect from information given to him by the contractor where quantities do not form part of the contract;

or

- by the quantity surveyor where quantities form part of the contract.

Under Clause 30.8 of the Standard Form of Building Contract, the architect is required to issue the Final Certificate as soon as possible but before the expiration of three months from either:

- the end of the Defects Liability Period stated in the Appendix (Clause 17.2)

or

- from completion of making good defects under Clause 17

or

- from receipt by either the architect or the quantity surveyor from the contractor of all the documents necessary to enable the adjustment of the contract sum including all documentation relating to the accounts of nominated sub-contractors or suppliers, whichever is the latest.

Under Clause 4.5 of the Intermediate Form of Contract the architect is required to send a copy of the computation of the adjusted contract sum (the Final Account) to the contractor within the Period of Final Measurement and Valuation (usually six months subsequent to the date of practical completion), prepared from information submitted by the contractor either before or within a reasonable time after practical completion. Clause 4.6 requires the architect to issue, within twenty-eight days of the date of sending the account, or the issue of a Certificate of Making Good Defects (see Clause 2.10), a Final Certificate certifying the amount due to the contractor in full and final settlement of the contract.

Under Clause 4.4 of the Agreement for Minor Building Works the contractor must supply within the period inserted in the contract clause (usually three months unless varied) all documentation necessary for the calculation of the amount to be finally certified by the architect. Providing that the architect has agreed the satisfactory completion of defects and issued a certificate as required by Clause 2.5, within twenty-eight days of the receipt of such documentation the architect must issue a final certificate stating the balance of the total value of works done under the contract.

It is therefore a contractual responsibility for the architect to prepare in some way an account showing:

- the adjustments due to alteration of precise sums included in the contract for specific items (ie provisional and P.C. sums);
- adjustment of measured works in contracts where quantities form part of the contract;
- adjustment of contract (variations), works agreed either as lump sums (accepted estimates or otherwise) or by daywork charges;
- increased cost claims in respect of wages and materials.

The agreement of the sums and amounts claimed may take some time but, when agreed in principle by the contractor, should be prepared, assembled and presented in a document known as the Final Account.

The presentation of the Final Account is a matter both of psychology and of tact. The account should be agreed by both parties to the contract – the employer and the contractor – and as the final cost has already been agreed with the contractor his formal signature on the account is all that is required (see Fig 7.3).

With regard to the client, matters can be somewhat difficult. If the final value of work is within the original contract sum few difficulties arise and his signature is usually quickly obtained. If the contract sum has been exceeded there may be some reluctance to accept the negotiated amount even where the employer may be aware that an overspending has occurred.

Methods are shown which illustrate the techniques applicable to the presentation of the same summary of account in both situations.

Method A (see Fig 7.1) comprises separate pages showing:

Page One: A summary sheet showing amount due by collection from which are deducted sums previously certified in Interim Certificates.

Page Two: Summary of variations, adjustment of provisional and P.C. sums.

Page Three: Summary of variations (including adjustment of measured rates where applicable and dayworks).

Page Four: Increased costs claim sheet.

Fig 7.1 Method A

Page one

Final Account,
New Building, South Road, New Town
For A. Client & Co Ltd
Job Number 259 November 1985

Messrs A. B. Construction Co Ltd
49 High Street
New Town

Messrs Line & Wash, architects
The Muse
Olympia

Summary of account	£
To value of contract dated [*insert date*]	58 681.78
Less omission of PC and provisional sums	27 600.00
	31 081.78
Add value of works covered by PC and provisional sums	26 800.00
	57 881.78
	26 800.00
	57 881.78
Add value of variations included in Architect's Instructions numbers [*1 - 184*]	1 247.14
	59 128.92
Add increased costs: Labour	1 014.41
Materials	699.27
	60 842.60

Signed: I. Jones RIBA
For Line & Wash, architects

Page two

Summary of variations, adjustment of PC and provisional sums

	Omissions £	Additions £
Contingency sum (item -)	1 500.00	
Structural steelwork (item -)	18 000.00	18 200.00
Heating and plumbing installation	4 800.00	5 030.00
Shutter doors	1 500.00	1 627.00
Electrical installation	1 800.00	1 943.00
Carried to account summary	27 600.00	26 800.00

Page three

Summary of variations included in Architect's Instructions numbers [*1 - 184*]

Instruction number	Omissions	Additions
[27] Adjustment of stormwater drainage		59.27
[34] Use emulsion paint in lieu of gloss to lavatory walls and ceilings etc	390.37	134.19
	851.84	2 098.98
		851.84
Net value of variations carried to account summary		1 247.14

Page four

Increased costs claim

Labour increases on [*insert date*]
Labour increased from [*insert date*] to [*insert date*]

	£
Tradesmen......hours at......p/hour	654.27
Labourers......hours at......p/hour	360.14
Increased cost of labour carried to account summary	1 014.41

To adjustment in cost of materials

	Basic rates	Actual rates
	£	£
	1 106.43	1 805.70
		1 106.43
Increased cost of materials carried to account summary		699.27

Method B (see Fig 7.2) comprises a running collection of items all as Method A but arranged in a different manner to show more precisely the responsibility for extras and where the additional expenditure has gone. For example:

1 Variations are split into two sections:

- Where variations were deemed necessary owing to contractual emergency (e.g. increased depth of foundations, non-availability of specified materials, etc.)
- Extra work or variations ordered by the employer subsequent to commencement of work on site.

The employer can then see at a glance how much his change of mind has actually cost. If the extras were ordered against accepted estimates the blow may be somewhat softened.

2 Increased costs are included at the end of the account. Contingencies are not included in the contract sum for this purpose. If provision is to be made, a provisional sum specified for this eventuality may be included. This, however, is rarely done.

It is usual for the architect to arrange a meeting with the employer to discuss the Final Account (see Figs 7.4, 7.5 and 7.6). Few laymen understand the contractual computations involved and a meeting to explain these is generally necessary. In addition to the negotiated account, details of all variations, dayworks, increased costs and sub-contractors' accounts should be available for inspection by the employer if he so desires.

The issue of the Final Certificate (see p. 158) is conclusive evidence that the contract works have been completed and executed in accordance with the conditions of the contract. It is therefore essential to establish that these are fulfilled in their entirety.

The Final Certificate states:

- The total of the sums previously paid to the contractor.
- The adjusted contract sum agreed after taking into account all adjustment, variations and increased cost claims.
- The amount due to the contractor in full and final payment.

The balance is payable within fourteen days after issue in the same way that it is under Interim Certificates: this period of fourteen days applies only to JCT 80 and MW 80 contracts – for IFC 84 it is twenty-one days. At the same time as the issue of the Final Certificate, the architect shall inform each nominated sub-contractor of the date of its issue (Figs 7.7, 7.8 and 7.9).

Fig 7.2 Method B

Final Account, November 1985
New Building, South Road, New Town
For A. Client & Co Ltd
Job Number 259

Messrs A. B. Construction Co Ltd
49 High Street
New Town

Messrs Line & Wash, architects
The Muse
Olympia

Summary of account

To value of contract dated [*insert date*]		58 631.78
Less omission of PC and provisional sums		27 600.00
		31 081.78
Add Structural steelwork	18 200.00	
Heating and plumbing	5 030.00	
Shutter door	1 627.00	
Electrical installation	1 943.00	
	26 800.00	26 800.00
		57 881.78

Add Variations included in Architect's Instructions numbers [*1 - 184*]

Instruction no.	Omissions	Extras	
	£	£	
27 Adjustment to stormwater drain		59.27	
34 Use of emulsion paint in lieu of gloss	390.27	134.19	
	851.84	2 098.98	
		851.84	
		1 247.14	1 247.14
			59 128.92

Add Increased costs
Labour increases on [*insert dates*] to [*insert dates*]

Tradesmen ... hours at ... p/hr	654.27	
Labourers ... hours at ... p/hr	360.14	
Materials	699.27	
	1 713.68	1 713.68
Final amount of contract		60 842.60

Signed: I. Jones
For Line & Wash, architects

Fig 7.3 Letter from architect to contractor: agreement to Final Account

Dear Sir

I enclose the statement of Final Account agreed for this contract and I should be obliged if you would kindly check this and, if you agree with the computation, sign where indicated and return to me at your earliest convenience.

I enclose a copy of the statement for your retention.

Yours faithfully

Under the Intermediate Form of Contract the amount certified as due to the contractor under the Final Certificate is payable within 21 days of the date of issue (Clause 4.6).

Under the Agreement for Minor Building Works the amount due must be paid within 14 days of the date of issue (Clause 4.4).

The form illustrated is applicable for use under all three forms of contract, the period for discharge being amended for the Intermediate Form of Contract from '14th' to '21st'.

Fig 7.4 Final Account

New Building
South Road
New Town

for A. Client & Co Ltd

Job Number 259

Messrs A. B. Construction Co Ltd
49 High Street
New Town

Messrs Line & Wash, architects
The Muse
Olympia

November 1976

Statement

To final amount of contract as Final Account attached	60 842.60
Less previous payments on account (certificates numbered 1 - 6)	56 000.00
Amount due in full and final settlement	4 842.60

We, having produced all relevant information, hereby certify that we have examined and agreed this statement and the accounts attached hereto and have no further claims under this contract.

Signed: I. Patcham
For and on behalf of
A. B. Construction Co Ltd

Fig 7.5 Letter from architect to client: to arrange meeting regarding Final Account.

Dear Sir

I have received all information and documentation from the contractor and after checking the sums claimed and amending these where necessary, have agreed the Final Account. The statement of Final Account has been signed by the contractor.

I should be pleased if you could let me know a date and time convenient to you to call at my office to discuss the amount.

Yours faithfully

Fig 7.6 Letter from architect to client: agreement to Final Account

Dear Sir

Thank you for your letter of [*insert date*] returning to me the statement of Final Account duly agreed and signed.

I have today issued my Final Certificate to the contractor in respect of the works for the sum of [*insert sum*] in full and final settlement of all financial obligation in respect of this contract and I enclose the certificate for your attention and discharge within fourteen days of its date.

The Final Certificate will be forwarded to you for settlement in the next few days.

Yours faithfully

Issued by:
address:

Final Certificate

Employer:
address:

Serial no:

Contractor:
address:

Job reference:

Issue date:

Works:
Situated at:

Original to Employer

Contract dated:

Under the terms of the above mentioned Contract,
the Contract Sum adjusted as necessary is . . . . £

The total amount previously certified for payment to the contractor is . . £

The difference between the above stated amounts is . . . . £

(in words)____

*Delete as appropriate

and is hereby certified as a balance due* to the Contractor from the Employer/*to the Employer from the Contractor.[1]

All amounts are exclusive of VAT.

To be signed by or for the issuer named above.

Signed____

*Delete as appropriate

1 The terms of the Contract provide that the amount shall as from the *14th/21st day after the date of this Certificate be a debt payable from the one to the other subject to any amounts properly deductible by the Employer.

This is not a Tax Invoice

Note: [1] Payment becomes due 14 days after issue where the contract is JCT 80 or MW 80 and 21 days after issue for IFC 84.

Fig 7.7 Letter from architect to contractor: issue of Final Certificate

Dear Sir

The Statement of Final Account has been signed by the employer and I have pleasure in enclosing a copy of my Final Certificate in respect of the contract.

The documentation submitted by you in support of the account is being returned to you under separate cover.

Yours faithfully

Fig 7.8 Letter from architect to client: to enclose statement of fees and expenses

Dear Sir

As requested I enclose a statement of my fees for professional services and out of pocket expenses, prepared in accordance with the Conditions of Engagement in respect of this contract, which I trust you will find in order.

Yours faithfully

Fig 7.9 Letter from architect to contractor: expression of thanks

Dear Sir

I have been asked by the employer to express to you his thanks for the efficient and prompt completion of the contract works. He is most satisfied with the results.

Yours faithfully

Glossary of Terms

Acceptance (of a tender) agreement to accept the tender or offer made by a contractor

Agent contractor's representative

Antiquities items or artifacts usually of medieval or earlier date

Architect person registered under the Architects (Registration) Acts 1931 to 1969

Architect's Appointment see 'Conditions of engagement'

Architect's Instruction direction or order from the architect to the contractor

Articles of agreement see 'Contract form'

Assignment (of agreements) the formal making over, or transferral

Bankruptcy insolvency

Basic rates market prices of materials or goods and national wages rates incorporated in a Schedule of Rates current at the date of tender

Bill of quantities schedule prepared in accordance with a standard method to show each type of work required for a particular building in recognised units of measurement

Block plan generally to a scale of 1:1250 to show adjoining streets, boundaries of site, size and position of proposed and existing buildings and other particulars as required under the Building Regulations

Brief summary of instructions

Building Regulations statutory standards for building works applicable to England and Wales except for the area administered under the London Building Acts

Cash flow movement of money through a firm's accounts

Certificate (Final) official notification issued by the architect advising both the employer and the contractor of the final agreed value of work executed and issued at the completion of the contract

Certificate (Interim) official notification issued, usually at monthly intervals, by the architect advising both the employer and the contractor of the value of work executed

Clerk of works person, usually a former building tradesman, appointed and paid by the employer to superintend the contract on his behalf

Client person employing the services of a professional consultant

Commission agency to act on behalf of another

Completion date date agreed at the beginning of a contract when the works will be practically completed for occupation by the employer

Completion (of works) standard of work executed when the contract has progressed sufficiently to enable occupation by the employer

Compliance (of contractual duties) agreement to comply

Composition (with creditors) by which a debtor on payment of an agreed proportion is freed from his obligations

Conditions of Engagement terms of employment of an architect by his client published by the RIBA as Architect's Appointment

Contract form printed document comprising a number of clauses defining the terms and conditions of the contractual relationship between employer and contractor, usually either the Standard Form of Building Contract, Intermediate Form of Contract, or the Agreement for Minor Building Works

Contract period period of time within which the contractor undertakes practically to complete the works

Contract (under hand) contract completed with the holograph signature of both parties

Contract (under seal) contract embossed with the seal of one or both of the contracting parties

Contractor builder who generally works under contract

Contingency sum monies included in a contract to cover unforeseen events or contingencies

Creditor one to whom money is owed

Datum known or assumed basis for reference

Daywork charges net cost of labour, materials and plant plus a percentage to cover administration costs, which percentage is agreed and included in the Appendix to the Standard Form of Building Contract

Default failure to meet an obligation

Defects Liability Period period of time included in the contract during which the contractor is liable to remedy specific faults at his sole cost

Determination (of contract) termination of a contractual situation

Employer joint signatory of the contract with the contractor, eventually the owner of the building

Estimate (approximate) rough guide costing generally prepared on the basis of a unit cost per square metre or on approximate quantities

Estimate (final) more exact estimate usually prepared by the quantity surveyor on the basis of priced quantities

Estimate (lump sum) estimate submitted by a contractor for specific work in which details of the build-up of the price are not included

Expenses (out of pocket) as defined in Part 4 of the Architect's Appointment published by the RIBA these are applicable to all architect's work

Extension (of time) extension of the contract period owing to factors closely defined in the contract form

Fee percentages scale of fees laid down in Part 4 of the Architect's Appointment published by the RIBA, applicable to most architect's commissions

Final Account priced summary of the eventual cost of the contract works

Foreman site representative of the contractor on small works contracts, usually a former craft tradesman

Form of Tender form used by most architect's practices for the standardisation of tenders

Instruction see 'Architect's Instruction'

Insurance (of contract works) in consideration of the payment of a sum of money or premium, the provision of cover against loss or damage to the works against specific contingencies

Key plan generally to a scale of 1:2500, provided when the location of the proposed building site is not clearly identifiable from the block plan.

Level reference point with a known height above a specific datum

Liquidator official appointed specifically to wind up a company

Make good to repair a portion of the structure or finishings to match exactly the surrounding work

Manager (contract) usually appointed on large contracts by the contractor to deal with contract management as opposed to craft practice

Measured work work of which the content can be measured and valued against agreed or known unit rates of cost

Moiety one half

Nominated (sub-contractors) firms or individuals who are appointed by the architect through a sub-contract with the contractor to carry out specific work or services as part of the contract

Nominated (suppliers) as above but to supply specific goods for the contract

Notification (Form of) advice to the nominated sub-contractor of the amount included in the certificate issued by the architect

Open up to expose or break out the structure for inspection or testing

Plant generally scaffolding and mechanical equipment

Prime cost sum basic costs before the addition of overhead charges and profit, usually incorporated in a specification or bill of quantities to cover work outside the normal sphere of a contractor's experience, or for materials or fittings selection of which cannot be made prior to the placing of the contract

Production drawings drawings prepared for the contract works

Provisional sum sum included in a specification or bill of quantities either for work carried out by statutory authorities or for work of which the extent or description cannot be properly described prior to the placing of the contract

Quantities (provisional) items of measurement of which the accuracy cannot be exactly quantified are annotated by the description 'provisional'. Subsequently the work is measured, the quantities are adjusted and the work revalued against the rates inserted against the items in the bills of quantities

Quantity surveyor professional consultant whose chief work is the preparation of bills of quantities

Rates unit cost of a measured item in a bill of quantities

Receiver person appointed to administer a debtor's property

Retention usually a sum of money held back from the contractor's interim payments to cover possible defects in the executed work

Schedule information in a format useful as a supplement to specifications and bills of quantities to locate specific finishes and components

Setting out careful laying out of dimensioned lines or profiles on site to delineate the outline of a building

Site meeting meeting for the purpose of discussion and decision on contractual matters

Sketch plan preliminary drawings prepared by an architect to illustrate his preliminary proposals for the project

Specification document prepared to convey to a contractor all information concerning a proposed building which cannot be easily shown on the architect's production drawings

Starting date date on which work on a contract commences, usually the date when possession of the site is given to the contractor

Statutory obligations obligations enforceable by law, either nationally or by local by-law; also regulations applicable to statutory undertakers eg. water, gas, electricity

Sub-contractor individual or firm appointed by the contractor to carry out contract work on his behalf (see also 'Nominated sub-contractors')

Sub-letting generally the appointment by the contractor of domestic sub-contractors who have not been nominated by the architect

Surety pledge of security, often the deposition of a sum of money

Suspension (of works) cessation

Temporary buildings site huts including offices and stores

Tender offer by a contractor to carry out specific work for a named sum of money

Tools generally hand implements

Trustee person placed in possession of a property who is legally obliged to administer it solely for the purpose specified

Variation any deviation from the specified contract works in scope, type of work, or material

Winding up order operation of ceasing trading, realising assets and discharging liabilities either voluntarily or by order of a court